Synthesis Lectures on Mechanical Engineering

This series publishes short books in mechanical engineering (ME), the engineering branch that combines engineering, physics and mathematics principles with materials science to design, analyze, manufacture, and maintain mechanical systems. It involves the production and usage of heat and mechanical power for the design, production and operation of machines and tools. This series publishes within all areas of ME and follows the ASME technical division categories.

Yu Guo · Albert C. J. Luo

Periodic Motions to Chaos in a Spring-Pendulum System

 Springer

Yu Guo
McCoy School of Engineering
Midwestern State University
Wichita Falls, TX, USA

Albert C. J. Luo
Department of Mechanical and Mechatronics
Engineering
Southern Illinois University Edwardsville
Edwardsville, IL, USA

ISSN 2573-3168   ISSN 2573-3176   (electronic)
Synthesis Lectures on Mechanical Engineering
ISBN 978-3-031-17885-6   ISBN 978-3-031-17883-2   (eBook)
https://doi.org/10.1007/978-3-031-17883-2

一夢醒來多少舟
爭流千弓舟
狎鷗風雨去書
頸花群坐主留
君知否莫讓年
莫付水流　朝俊

Preface

In 1933, such a spring-pendulum system was proposed by Witt and Gorelik as a two-degree-of-freedom parametric oscillator, and the spring-pendulum system possesses an internal parametric resonant system. Such a system was used for explanation of the internal resonance phenomena. Traditionally, the Taylor series was used for expanding sinusoidal function of the pendulum, and the perturbation method was employed for approximate dynamical behaviors in the spring-pendulum. The linear results were extensively used to explain physical phenomena. However, such an approximated spring-pendulum cannot provide accurate solutions and cannot adequately explain the real physical phenomena as well. Thus, this book presents accurate, semi-analytical solutions of the non-approximated spring-pendulum, and the bifurcation trees of periodic motions to chaos in such a nonlinear spring-pendulum are presented. The harmonic frequency-amplitude characteristics of periodic motions to chaos are presented. However, the semi-analytical results of periodic motions for the nonlinear spring-pendulum cannot be obtained through perturbation methods.

In this book, periodic motions to chaos in a nonlinear spring-pendulum are discussed. In Chap. 2, the implicit mapping method is presented as a semi-analytical method for periodic motions in nonlinear dynamical systems. In Chap. 3, the mathematical modeling of a nonlinear spring-pendulum is developed, and the corresponding discretization of the non-approximated differential equations is completed for the implicit discrete maps. In Chap. 4, periodic motions in the periodically forced nonlinear spring-pendulum are studied through the mapping structures, and the finite-Fourier series is also presented for periodic motions. The bifurcation tree of period-1 motions to chaos varying with excitation frequency is presented in Chap. 5. In Chap. 6, higher-order periodic motions to chaos varying with excitation amplitude are discussed through period-3 motion to chaos.

Finally, the authors hope the materials presented herein can last long for science and engineering.

Wichita Falls, TX, USA
Edwardsville, IL, USA
Yu Guo
Albert C. J. Luo

Contents

About the Authors

Yu Guo is Associate Professor at Midwest State University. His research lies in nonlinear dynamical systems, discontinuous dynamics, nonlinear dynamical systems control. He is an associate editor for *Journal of Vibration Testing and Systems Dynamics.*

Albert C. J. Luo is Distinguished Research Professor at Southern Illinois University Edwardsville. He is an international recognized scientist on nonlinear dynamics, discontinuous dynamical systems. His main contributions are on a local singularity theory for discontinuous dynamical systems, dynamical systems synchronization, generalized harmonic balance method for analytical solutions of periodic motions, implicit mapping method for semi-analytical solutions of periodic motions.

Introduction 1

The spring-pendulum system possesses an internal parametric resonant system. One used such a system to investigate the internal resonance phenomena. Traditionally, one used the Taylor series to expand sinesoidal function of the pendulum, and the perturbation method was adopted to approximately determine dynamical behaviors in the spring-pendulum. The linear results were extensively used to explain physical phenomena. However, such linearized spring-pendulum cannot provide accurate solutions and cannot adequately explain the real physical phenomena as well.

The early studies on periodic motion in nonlinear systems were mainly based on the perturbation methods. In 1788, Lagrange [1] investigated the three-body problems as a perturbation of the two-body problems. In 1899, Poincare [2] developed the perturbation theory for the periodic motions of celestial bodies. In 1920, van der Pol [3] studied the periodic solutions of an oscillator circuit through the method of averaging. In 1928, Fatou [4] proved the asymptotic validity of the method of averaging based on the solution existence theorems of differential equations. In 1935, Krylov and Bogoliubov [5] extended the method of averaging to nonlinear oscillations in nonlinear vibration systems. In 1964, Hayashi [6] presented the perturbation methods including averaging method and the principle of harmonic balance method for nonlinear oscillations. In 1969, Barkham and Soudack [7] employed the extended Krylov–Bogoliubov method for the approximate solutions of a second-order nonlinear autonomous differential equations. In 1987, Garcia-Margallo and Bejarano [8] used a generalized harmonic balance method for the approximate solutions of nonlinear oscillations. Yuste and Bejarano [9, 10] used the elliptic functions instead of trigonometric functions to improve the Krylov–Bogoliubov method on nonlinear oscillators. In 1990, Coppola and Rand [11] obtained the approximation of limit cycles through the method of averaging via elliptic functions.

Y. Guo and A. C. J. Luo, *Periodic Motions to Chaos in a Spring-Pendulum System*,
Synthesis Lectures on Mechanical Engineering,
https://doi.org/10.1007/978-3-031-17883-2_1

In 1933, such a spring pendulum system was proposed in Witt and Gorelik [12] as a two-degree-of-freedom parametric oscillator. In 1976, Olson [13] discussed the autoparametric resonance of a spring pendulum using the harmonic balance method. In 1978, Falk [14] studied the recurrence effects of a parametric spring pendulum using the perturbation method. In 1984, Lai [15] investigated the recurrence of a resonant spring pendulum system; and the trajectories of motions are correlated with experiments. In 1990, numerical investigations on chaos in a spring pendulum system without damping were presented (e.g., Nunez-Yepez et al. [16], Cuerno et al. [17]). In 1993, Bayly and Virgin [18] used the Taylor series to expand the sinusoidal function up to the cubic terms, and the periodic motion was assumed to the first-order harmonic term with constant for (1:2)-internal resonance. In addition, the experimental studies were completed. In 1994, Lee and Hsu [19] used harmonic balance method to obtain the approximate stead-state solutions of a spring pendulum; and the global domains of attraction for the periodic motions were presented using cell-to-cell mapping method. In 1996, Weele and Kleine [20] presented a detailed order-chaos-order sequence of such a spring pendulum system. In 1999, Lee and Park [21] proposed a second-order approximation for the chaotic responses in a harmonically excited spring pendulum. In 2003, Eissa et al. [22] studied the resonance and stability of a non-linear spring pendulum under harmonic excitation and Popov [23] used Poincare maps to investigate the chaotic motions of a spring pendulum, providing an insight to nonlinear shell vibrations. In 2006, Alasty and Shabani [24] studied the chaotic responses of a spring pendulum system through bifurcation diagram and Poincare maps. In 2008, Eissa et al. [25] studied the vibration reduction of a nonlinear spring pendulum through the perturbation method. In 2009, Amer and Bek [26] investigated the chaotic responses of a harmonically excited spring pendulum moving in a circular path. In 2011, Markeyev [27] studied the rotating motions and the stability in the spring pendulum using the perturbation method. In 2018, Sousa et al. [28] studied the energy distributions of a spring pendulum system. However, the traditional perturbation method does not provide an effective route towards analytical predictions of the nonlinear dynamical behaviors of such a system.

On the other hand, researchers have been interested in analytical predictions of periodic motions for nonlinear oscillatory systems. In 2012, Luo [29] developed an analytical method for analytical solutions of periodic motions in nonlinear dynamical systems, which was based on the generalized harmonic balance. The detailed discussion can be found in Luo [30–32]. Luo and Huang [33] used the generalized harmonic balance method for the analytical solutions of period-1 motions in the Duffing oscillator with a twin-well potential. Luo and Huang [34] also employed a generalized harmonic balance method to find analytical solutions of period-m motions in such a Duffing oscillator. The analytical bifurcation trees of periodic motions to chaos in the Duffing oscillator were obtained (also see, Luo and Huang [35–40]). Such analytical bifurcation trees showed the connection from periodic motions to chaos analytically. Luo and Yu [41] also employed the generalized harmonic balance method for approximated analytical solutions of period-1 motions in a nonlinear quadratic oscillator. Analytical solutions of periodic motions in van der Pol oscillator were also presented by Luo and Laken [42, 43]. The analytical solutions of period-m motions to chaos in the van der Pol-Duffing oscillator were studied in Luo and Laken [44]. In 2016, Luo and Yu [45, 46] discussed the analytical solutions for the bifurcation trees of period-1 motions to chaos in a two-degree-of-freedom nonlinear oscillator. In 2013, Luo [47] extended such ideas of the generalized harmonic method for periodic motions in time-delay, nonlinear dynamical systems. Luo and Jin [48] applied such a method for the bifurcation tree of period-1 motion to chaos in a periodically forced, quadratic nonlinear oscillator with time-delay. Further, Luo and Jin [49–51] investigated the periodic motions to chaos in a time-delayed Duffing oscillator. The generalized harmonic balance method is suitable for periodic motions in nonlinear systems with polynomial nonlinearity.

However, such a method is difficult to be applied to nonlinear dynamical systems with non-polynomial functions (e.g., pendulum system). Thus, in 2015, Luo [52] developed a semi-analytical method for periodic motions in complicated nonlinear dynamical systems. Luo and Guo [53, 54] applied such a method to predict the bifurcation trees of periodic motions in a Duffing oscillator. Luo and Xing [55, 56] employed the method for symmetric and asymmetric period-1 motions and bifurcation trees of period-1 motions to chaos in a delayed, hardening Duffing oscillator. Luo and Xing [57] investigated the time-delay effects for periodic motions in a time-delay, Duffing oscillator. Furthermore, Xing and Luo [58, 59] subsequently studied the periodic motions in a twin-well, Duffing oscillator with time-delay displacement feedback and discussed the possibility of infinity bifurcation trees in such a nonlinear oscillator. To study higher order dynamical systems, Xu and Luo [60] applied this method for a coupled van der Pol Duffing oscillator and discovered a series of periodic motions. Xu and Luo [61] also presented sequential period-$(2m - 1)$ motions to chaos in the periodically forced van der Pol oscillator. In 2016, Guo and Luo [62, 63] employed such a method on a periodically forced pendulum. The vibrational and rotational periodic motions to chaos in the periodically forced pendulum were obtained. To know the inherent complex dynamics of the parametrically excited pendulum, Guo and Luo [64] investigated a parametrically excited pendulum system through the implicit mapping method. The parametric pendulum possesses different dynamical behaviors in compared to the periodically forced pendulum. To help one understand dynamics of the parametric pendulum, the bifurcation trees of periodic motions in the parametrically excited pendulum was presented. In 2020, Luo and Yuan [65] investigated a periodically excited nonlinear spring-pendulum, and the bifurcation trees with varying excitation frequency in such a nonlinear spring-pendulum were presented. To considered excitation amplitude effects on periodic motion dynamics in the spring-pendulum, Guo and Luo [66] obtained the bifurcation trees of period-1 motion to chaos. Guo and Luo [67] studied the bifurcation dynamics of higher-order periodic motions to chaos for the nonlinear spring-pendulum. In engineering application, higher order periodic motions and corresponding bifurcations become very important in physical and engineering applications. One used the spring-pendulum for energy absorber for mechanical and civil structures [68] and for energy harvester to harvest multi-directional vibrational energy at ultra-low frequency [69]. The dynamics of a human running was modelled through the spring-pendulum [70]. As a typical two degree of freedom (2-DOF) oscillating system with nonlinear coupling, a spring pendulum system possesses very complicated dynamical behaviors, which has drawn research interests for many years.

Thus, in this book, periodic motions to chaos in a nonlinear spring-pendulum will be addressed. The implicit mapping method will be presented as a semi-analytical method for periodic motions in nonlinear dynamical systems in Chap. 2. The physical problem discretization of a nonlinear spring-pendulum will be placed in Chap. 3. In Chap. 4, periodic motions in the periodically forced nonlinear spring-pendulum will be studied from the mapping structures, and the finite-Fourier series will be also presented for periodic motions. The bifurcation trees of period-1 motions to chaos will be discussed in Chap. 5, and the harmonic frequency-amplitude characteristics will be presented for nonlinear dynamics of periodic motions in the nonlinear spring-pendulum. In Chap. 6, higher-order periodic motions to chaos will be discussed through period-3 motion to chaos. The bifurcation trees of periodic motions to chaos varying with excitation amplitudes will be given, and harmonic amplitudes will be presented to show harmonic term effects on periodic motions. Finally complex periodic motions in the nonlinear spring-pendulums will be presented to show periodic motion complexity.

References

1. Lagrange JL (1788) *Mecanique Analytique* (2 vol.) (edition Albert Balnchard: Paris, 1965).
2. Poincare H (1899) *Methodes Nouvelles de la Mecanique Celeste*, Vol. 3, Gauthier-Villars: Paris.
3. van der Pol B (1920) A theory of the amplitude of free and forced triode vibrations. *Radio Review*, **1**: 701–710, 754–762.
4. Fatou P (1928), Sur le mouvement d'un systeme soumis 'a des forces a courte periode. Bull. Soc. Math., **56**: 98–139.
5. Krylov NM, Bogolyubov NN (1935), Methodes approchees de la mecanique non-lineaire dans leurs application a l'Aeetude de la perturbation des mouvements periodiques de divers phenomenes de resonance s'y rapportant. Kiev: Academie des Sciences d'Ukraine (in French).
6. Hayashi C (1964) *Nonlinear oscillations in Physical Systems* (McGraw-Hill Book Company, New York).
7. Barkham PGD, Soudack AC (1969) An extension to the method of Krylov and Bogoliubov. *International Journal of Control*, **10**: 377–392.
8. Garcia-Margallo J, Bejarano JD (1987) A generalization of the method of harmonic balance. *Journal of Sound and Vibration*, **116**: 591–595.
9. Yuste SB, Bejarano JD (1986) Construction of approximate analytical solutions to a new class of non-linear oscillator equations. *Journal of Sound and Vibration*, **110(2)**: 347–350.
10. Yuste SB, Bejarano JD(1990) Improvement of a Krylov-Bogoliubov method that uses Jacobi elliptic functions. *Journal of Sound and Vibration*, **139(1)**: 151–163.

11. Coppola VT, Rand RH (1990) Averaging using elliptic functions: Approximation of limit cycle. *Acta Mechanica*, **81**:125–142.
12. Witt A, Gorelik G (1933) Kolebanya uprugogo mayatnika, kak primer kolebanii dvuh parametricheski svyazannykh lineinikh sistem, *Zh. Tekh. Fiz.*, **3**:294–306. (The oscillation of an elastic pendulum as an example of the oscillations of two parametrically connected linear systems. *Zh. Tekh. Fiz.*, **3**(2–3):294–306.)
13. Olsson MG (1976) Why does amass on a spring sometimes misbehave? *American Journal of Physics*, **44**(12):1211–1212.
14. Falk L(1978) Recurrence effects in the parametric spring pendulum. *American Journal of Physics* **46**(11):120–1123.
15. Lai HM (1984) On the recurrence phenomenon of a resonant spring pendulum. *American Journal of Physics*, **52**(3): 219–223.
16. Nunez-Yepez HN, Salas-Brito AL, Vargas CA, Vicente L (1990) Onset of chaos in an extensible pendulum, *Physics Letters A*, **145**:101–105.
17. Cuerno R, Ranada AF, Ruiz-Lorenzo JJ (1992) Deterministic chaos in the elastic pendulum: A simple laboratory for non-linear dynamics. *American Journal of Physics*, **60**(1): 73–79.
18. Bayly PV, Virgin LN (1993) An empirical investigation of the stability of periodic motion in the forced spring-pendulum. *Proceedings of the Royal Society of London*, **443A**:391–408.
19. Lee WK, Hsu CS (1994) A global analysis of a harmonically excited spring-pendulum system with internal resonance. *Journal of Sound and Vibration*, **171**:335–359.
20. van der Weele JP, de Kleine E (1996) The order-chaos-order sequence in the spring-pendulum. *Physica A*, **228**:245–272.
21. Lee W.K. and Park, H.D., 1999, Second-order approximation for chaotic responses of a harmonically excited spring-pendulum system, *International Journal of Nonlinear Mechanics*, **34**:749–757.
22. Eissa M, EL-Serafi SA, EL-Sheikh M, Sayed M (2003) Stability and primary simultaneous resonance of harmonically excited non-linear spring pendulum system. *Applied Mathematics and Computation*, **145**: 421–442
23. Popov AA (2003) The application of spring pendulum analogies to the understanding of non-linear shell vibration. *Computational Fluid and Solid Mechanics*, Elsevier, pp. 590–594.
24. Alasty A, Shabani R (2006) Chaotic motions and fractal Basin boundaries in spring pendulum system. *Nonlinear Analysis: Real World Applications*, **7**: pp. 81–95.
25. Eissa M, Sayed M (2008) Vibration reduction of a three DOF non-linear spring pendulum. *Communications in Nonlinear Science and Numerical Simulation*, **13**: 465–488.
26. Amer TS, Bek MA (2009) Chaotic responses of a harmonically excited spring pendulum moving in circular path. *Nonlinear Analysis: Real World Applications*, **10**: 3196–3202.
27. Markeyev AP(2011) A case of plane rotations of an elastic pendulum. Journal of Applied Mathematics and Mechanics, 75.501–507.
28. de Sousa MC, Marcus FA, Caldas IL, Viana RL (2018)Energy distribution in intrinsically coupled systems: The spring pendulum paradigm. *Physica A*, **509**:1110–1119.
29. Luo ACJ (2012) *Continuous Dynamical Systems*. HEP/L&H Scientific: Beijing/Glen Carbon.
30. Luo ACJ (2014), On analytical routes to chaos in nonlinear systems. *International Journal of Bifurcation and Chaos*, **24** (4), 1430013 (28 pages).
31. Luo ACJ (2014) Toward Analytical Chaos in Nonlinear Dynamical Systems. Wiley.
32. Luo ACJ (2014) Analytical Routes to Chaos in Nonlinear Engineering. Wiley.
33. Luo ACJ, Huang J (2012) Approximate solutions of periodic motions in nonlinear systems via a generalized harmonic balance. *Journal of Vibration and Control*, **18**: 1661–1671.

34. Luo ACJ, Huang JZ (2012) Analytical dynamics of period-m flows and chaos in nonlinear Systems. *International Journal of Bifurcation and Chaos*, **22**(4), 29 pages. (Article No. 1250093, 29).

35. Luo ACJ, Huang JZ (2012) Analytical routes of period-1 motions to chaos in a periodically forced Duffing oscillator with a twin-well potential. *Journal of Applied Nonlinear Dynamics*, **1**: 73–108.

36. Luo ACJ, Huang JZ (2012) Unstable and stable period-m motions in a twin-well potential Duffing oscillator. *Discontinuity, Nonlinearity and Complexity*, **1**: 113–145.

37. Luo ACJ, Huang JZ (2013) Analytical solutions for asymmetric periodic motions to chaos in a hardening Duffing oscillator. *Nonlinear Dynamics*, **72**, 417–438.

38. Luo ACJ, Huang JZ (2013) Analytical period-3 motions to chaos in a hardening Duffing oscillator. *Nonlinear Dynamics*, **73**: 1905–1932.

39. Luo ACJ, Huang JZ (2013) An analytical prediction of period-1 motions to chaos in a softening Duffing oscillator. *International Journal of Bifurcation and Chaos*, **23**(5): Article No: 1350086 (31 pages).

40. Luo ACJ, Huang JZ (2014) Period-3 motions to chaos in a softening Duffing oscillator. *International Journal of Bifurcation and Chaos*, **24**: Article 1430010 (26 pages).

41. Luo ACJ, Yu B (2013) Complex period-1 motions in a periodically forced, quadratic nonlinear oscillator. *Journal of Vibration of Control*, **21**(5): 907–918.

42. Luo ACJ, Laken AB (2013) Analytical solutions for period-*m* motions in a periodically forced van der Pol Oscillator. *International Journal of Dynamics and Control*, **1**(2): 99–155.

43. Luo ACJ, Laken AB (2014) An approximate solution for period-1 motions in a periodically forced van der Pol oscillator. *ASME Journal of Computational and Nonlinear Dynamics*, **9**(3): 031001 (7 pages).

44. Luo ACJ, Laken AB (2014) Period-m motions and bifurcation in a periodically forced van der Pol-Duffing oscillator. *International Journal of Dynamics and Control*, **2**: 474–493.

45. Luo ACJ, Yu B (2015) Bifurcation trees of period-1 motions to chaos in a two-degree-of-freedom, nonlinear oscillator. *International Journal of Bifurcation and Chaos*, **25**(13), Article No:1550179 (40 pages).

46. Yu B, Luo ACJ (2017) Analytical period-1 motions to chaos in a two-degree-of-freedom oscillator with a hardening nonlinear spring, *International Journal of Dynamics and Control*, **5**(3):436–453.

47. Luo ACJ (2013) Analytical solutions for periodic motions to chaos in nonlinear systems with/without time-delay, *International Journal of Dynamics and Control*, **1**, 330–359.

48. Luo ACJ, Jin HX (2014) Bifurcation trees of period-m motion to chaos in a Time-delayed, quadratic nonlinear oscillator under a periodic excitation. *Discontinuity, Nonlinearity, and Complexity*, **3**, 87–107.

49. Luo ACJ, Jin HX (2014) Period-*m* motions to chaos in a periodically forced Duffing oscillator with a time-delay feedback. *International Journal of Bifurcation and Chaos*, **24** (10): Article no: 1450126 (20 pages).

50. Luo ACJ, Jin HX (2015) Complex period-1 motions of a periodically forced Duffing oscillator with a time-delay feedback. *International Journal of Dynamics and Control*, **3**: 325–340.

51. Luo ACJ, Jin HX (2015) Period-3 motions to chaos in a periodically forced Duffing oscillator with a time-delay feedback. *International Journal of Dynamics and Control*, **3**: 371–388.

52. Luo ACJ (2015) Periodic flows to chaos based on discrete implicit mappings of continuous nonlinear systems. *International Journal of Bifurcation and Chaos*, **25**(3): Article No. 1550044 (62 pages).

53. Luo ACJ, Guo Y (2015) A semi-analytical prediction of periodic motions in Duffing oscillator through mapping structures: *Discontinuity, Nonlinearity, and Complexity*, **4**(2): 13–44.

54. Guo Y, Luo ACJ (2015) On complex periodic motions and bifurcations in a periodically forced, damped, hardening Duffing oscillator. *Chaos, Solitons and Fractals*, **81**: 378–399.

55. Luo ACJ, Xing SY (2016) Symmetric and asymmetric period-1motion in a periodically forced, time-delayed, hardening Duffing oscillator. *Nonlinear dynamics*, **85**(2), 1141–1161.

56. Luo ACJ, Xing SY (2016) Multiple bifurcation trees of period-1 motions to chaos in a periodically forced, time-delayed, hardening Duffing oscillator. *Chaos, Solitons & Fractals*, **89**, 405–434.

57. Luo ACJ, Xing SY (2017) Time-delay effects on periodic motions in a Duffing oscillator. In *Chaotic, Fractional, and Complex Dynamics: New Insights and Perspectives*, edited by Mark Edelman, Elbert E. N. Macau, and Miguel A. F. Sanjuan. Understanding Complex Systems. Cham: Springer.

58. Xing SY and Luo ACJ (2017) Towards infinite bifurcation trees of period-1 motions to chaos in a time-delayed, twin-well Duffing oscillator." *Journal of Vibration Testing and System Dynamics*, 1(4), 353–392.

59. Xing SY and Luo ACJ (2018) On possible infinite bifurcation trees of period-3 motions to chaos in a time-delayed, twin-well Duffing oscillator. *International Journal of Dynamics and Control*, **6**(4),1429–1464.

60. Xu YY, Luo ACJ (2019) A series of symmetric period-1 motions to chaos in a two-degree-of-freedom van der Pol-Duffing oscillator. *Journal of Vibration Testing and System Dynamics*, **2**(2), 119–153.

61. Xu YY, Luo ACJ (2019) Sequent period-(2m − 1) motions to chaos in the van del Pol oscillator. *International Journal of Dynamics and Control*, **7**(3), 795–807.

62. Guo Y, Luo ACJ (2016) Routes of periodic motions to chaos in a periodically forced pendulum. *International Journal of Dynamics and Control* **5**(3): 551–569.

63. Luo ACJ, Guo Y (2016) Periodic motions to chaos in pendulum. *International Journal of Bifurcation and Chaos* **26**(9): 1650159 (64 pages).

64. Guo Y, Luo ACJ (2017) Complete bifurcation trees of a parametrically driven pendulum. *Journal of Vibration Testing and System Dynamics*, **1**(2): 93–134.

65. Luo ACJ, Yuan YG (2020) Bifurcation trees of period-1 to period-2 motions in a periodically excited nonlinear spring pendulum. *Journal of Vibration Testing and System Dynamics*, **4**(3):201–248.

66. Guo Y (2022) Bifurcations and harmonic responses of period-1 motions in a periodically excited spring pendulum. *Journal of Vibration Testing and System Dynamics*, 6(3):297–315.

67. Guo Y, Luo ACJ (2022) Period-3 motions to chaos in a periodically forced nonlinear-spring pendulum. *AIP Chao*, **32**: 103129.

68. Anh ND, Matsuhisa H, Viet LD, Yasuda M (2007) Vibration control of an inverted pendulum type structure by passive mass-spring-pendulum dynamic vibration absorber. *Journal of Sound and Vibration*, **307**:187–201.

69. Wu YP, Qiu JH, Zhou SP, Ji HL, Chen Y, Li S (2018) A piezoelectric spring pendulum oscillator used for multi-directional and ultra-low frequency vibration energy harvesting. *Applied Energy*, **231**: 600–614.

70. Masters SE, Challis JH (2021) Increasing the stability of the spring loaded inverted pendulum model of running with a wobbling mass. *Journal of Biomechanics*, **123**: 110527.

A Semi-analytical Method **2**

From Luo [1, 2], the implicit mapping method for nonlinear dynamical systems is presented through the following theorem.

Theorem 1 *Consider a nonlinear dynamical system as*

$$\dot{\mathbf{x}} = \mathbf{f}(\mathbf{x}, t, \mathbf{p}) \in \mathbf{R}^n \tag{2.1}$$

where $\mathbf{f}(\mathbf{x}, t, \mathbf{p})$ is a C^r-continuous nonlinear vector function $(r \geq 1)$. If such a dynamical system has a periodic motion $\mathbf{x}(t)$ with finite norm $||\mathbf{x}||$ and period $T = 2\pi/\Omega$, there is a set of discrete time t_k $(k = 0, 1, \ldots, N)$ with $(N \to \infty)$ during one period T, and the corresponding solution $\mathbf{x}(t_k)$ and vector fields $\mathbf{f}(\mathbf{x}(t_k), t_k, \mathbf{p})$ are exact. Suppose a discrete node $\mathbf{x}_k$ is on the approximate solution of the periodic motion under $||\mathbf{x}(t_k) - \mathbf{x}_k|| \leq \varepsilon_k$ with a small $\varepsilon_k \geq 0$ and

$$||\mathbf{f}(\mathbf{x}(t_k), t_k, \mathbf{p}) - \mathbf{f}(\mathbf{x}_k, t_k, \mathbf{p})|| \leq \delta_k \tag{2.2}$$

where a small $\delta_k \geq 0$. During a time interval $t \in [t_{k-1}, t_k]$, there is a general implicit mapping $P_k : \mathbf{x}_{k-1} \to \mathbf{x}_k$ $(k = 1, 2, \ldots, N)$ as

$$\mathbf{x}_k = P_k \mathbf{x}_{k-1} \quad \text{with } \mathbf{g}_k(\mathbf{x}_{k-1}, \mathbf{x}_k, \mathbf{p}) = \mathbf{0}, \quad k = 1, 2, \ldots, N. \tag{2.3}$$

where $\mathbf{g}_k$ is an implicit vector function. Consider a mapping structure as

$$P = P_N \circ P_{N-1} \circ \cdots \circ P_2 \circ P_1 : \mathbf{x}_0 \to \mathbf{x}_N;$$
$$\text{with } P_k : \mathbf{x}_{k-1} \to \mathbf{x}_k \quad (k = 1, 2, \ldots, N). \tag{2.4}$$

For $\mathbf{x}_N = P\mathbf{x}_0$, if there is a set of nodes points $\mathbf{x}_k^*$ $(k = 0, 1, \ldots, N)$ computed by

© The Author(s), under exclusive license to Springer Nature Switzerland AG 2023   9
Y. Guo and A. C. J. Luo, *Periodic Motions to Chaos in a Spring-Pendulum System*,
Synthesis Lectures on Mechanical Engineering,
https://doi.org/10.1007/978-3-031-17883-2_2

$$\mathbf{g}_k(\mathbf{x}_{k-1}^*, \mathbf{x}_k^*, \mathbf{p}) = \mathbf{0}, \quad (k = 1, 2, \ldots, N)$$
$$\mathbf{x}_0^* = \mathbf{x}_N^*, \tag{2.5}$$

then the points $\mathbf{x}_k^*$ *(*$k = 0, 1, \ldots, N$*) are approximations of points* $\mathbf{x}(t_k)$ *of the periodic solution. In a neighborhood of* $\mathbf{x}_k^*$*, with* $\mathbf{x}_k = \mathbf{x}_k^* + \Delta\mathbf{x}_k$*, the linearized equation is given by*

$$\Delta\mathbf{x}_k = DP_k \cdot \Delta\mathbf{x}_{k-1}$$
$$\text{with } \mathbf{g}_k(\mathbf{x}_{k-1}^* + \Delta\mathbf{x}_{k-1}, \mathbf{x}_k^* + \Delta\mathbf{x}_k, \mathbf{p}) = \mathbf{0}$$
$$(k = 1, 2, \ldots, N). \tag{2.6}$$

The resultant Jacobian matrix of the periodic motion is

$$DP_{k(k-1)\cdots 1} = DP_k \cdot DP_{k-1} \cdot \ldots \cdot DP_1, \quad (k = 1, 2, \ldots, N);$$
$$DP \equiv DP_{N(N-1)\cdots 1} = DP_N \cdot DP_{N-1} \cdot \ldots \cdot DP_1 \tag{2.7}$$

where

$$DP_k = \left[\frac{\partial \mathbf{x}_k}{\partial \mathbf{x}_{k-1}}\right]_{(\mathbf{x}_{k-1}^*, \mathbf{x}_k^*)} = -\left[\frac{\partial \mathbf{g}_k}{\partial \mathbf{x}_k}\right]_{(\mathbf{x}_{k-1}^*, \mathbf{x}_k^*)}^{-1} \left[\frac{\partial \mathbf{g}_k}{\partial \mathbf{x}_{k-1}}\right]_{(\mathbf{x}_{k-1}^*, \mathbf{x}_k^*)}. \tag{2.8}$$

The eigenvalues of DP *and* $DP_{k(k-1)\cdots 1}$ *for such a periodic motion are determined by*

$$|DP_{k(k-1)\cdots 1} - \overline{\lambda}\mathbf{I}_{n\times n}| = 0, \quad (k = 1, 2, \ldots, N);$$
$$|DP - \lambda\mathbf{I}_{n\times n}| = 0. \tag{2.9}$$

Thus, the stability and bifurcation of the periodic motion can be classified by the eigenvalues of $DP(\mathbf{x}_0^*)$ *with*

$$([n_1^m, n_1^o] : [n_2^m, n_2^o] : [n_3, \kappa_3] : [n_4, \kappa_4] | n_5 : n_6 : [n_7, l, \kappa_7]) \tag{2.10}$$

where n_1 *is the total number of real eigenvalues with magnitudes less than one (*$n_1 = n_1^m + n_1^o$*),* n_2 *is the total number of real eigenvalues with magnitude greater than one (*$n_2 = n_2^m + n_2^o$*),* n_3 *is the total number of real eigenvalues equal to* $+1$*;* n_4 *is the total number of real eigenvalues equal to* -1*;* n_5 *is the total pair number of complex eigenvalues with magnitudes less than one,* n_6 *is the total pair number of complex eigenvalues with magnitudes greater than one,* n_7 *is the total pair number of complex eigenvalues with magnitudes equal to one.*

(i) *If the magnitudes of all eigenvalues of* DP *are less than one, the approximate period-1 motion is stable.*

(ii) *If at least the magnitude of one eigenvalue of* DP *is greater than one, the approximate period-1 motion is unstable.*

(iii) *The boundaries between stable and unstable periodic motion give bifurcation and stability conditions.*

The period-m motion in a nonlinear dynamical system can be described through $(mN+1)$ nodes for period mT. As in Luo [1, 2], the corresponding theorem is presented as follows.

Theorem 2 *Consider a nonlinear dynamical system in Eq. (2.1). If such a dynamical system has a period-m motion $\mathbf{x}^{(m)}(t)$ with finite norm $||\mathbf{x}^{(m)}||$ and period mT $(T = 2\pi/\Omega)$, there is a set of discrete time t_k $(k = 0, 1, \ldots, mN)$ with $(N \to \infty)$ during m-periods (mT), and the corresponding solution $\mathbf{x}^{(m)}(t_k)$ and vector field $\mathbf{f}(\mathbf{x}^{(m)}(t_k), t_k, \mathbf{p})$ are exact. Suppose discrete node $\mathbf{x}_k^{(m)}$ is on the approximate solution of the periodic motion under $||\mathbf{x}^{(m)}(t_k) - \mathbf{x}_k^{(m)}|| \le \varepsilon_k$ with a small $\varepsilon_k \ge 0$ and*

$$||\mathbf{f}(\mathbf{x}^{(m)}(t_k), t_k, \mathbf{p}) - \mathbf{f}(\mathbf{x}_k^{(m)}, t_k, \mathbf{p})|| \le \delta_k \tag{2.11}$$

with a small $\delta_k \ge 0$. During a time interval $t \in [t_{k-1}, t_k]$, there is a general implicit mapping $P_k : \mathbf{x}_{k-1}^{(m)} \to \mathbf{x}_k^{(m)}$ $(k = 1, 2, \ldots, mN)$ as

$$\mathbf{x}_k^{(m)} = P_k \mathbf{x}_{k-1}^{(m)} \quad \text{with } \mathbf{g}_k(\mathbf{x}_{k-1}^{(m)}, \mathbf{x}_k^{(m)}, \mathbf{p}) = \mathbf{0}, \quad k = 1, 2, \ldots, mN \tag{2.12}$$

where $\mathbf{g}_k$ is an implicit vector function. Consider a mapping structure as

$$P = P_{mN} \circ P_{mN-1} \circ \cdots \circ P_2 \circ P_1 : \mathbf{x}_0^{(m)} \to \mathbf{x}_{mN}^{(m)};$$
$$\text{with } P_k : \mathbf{x}_{k-1}^{(m)} \to \mathbf{x}_k^{(m)} \quad (k = 1, 2, \ldots, mN). \tag{2.13}$$

For $\mathbf{x}_{mN}^{(m)} = P\mathbf{x}_0^{(m)}$, if there is a set of points $\mathbf{x}_k^{(m)*}$ $(k = 0, 1, \ldots, mN)$ computed by

$$\mathbf{g}_k(\mathbf{x}_{k-1}^{(m)*}, \mathbf{x}_k^{(m)*}, \mathbf{p}) = \mathbf{0}, \quad (k = 1, 2, \ldots, mN)$$
$$\mathbf{x}_0^{(m)*} = \mathbf{x}_{mN}^{(m)*}, \tag{2.14}$$

then the points $\mathbf{x}_k^{(m)*}$ $(k = 0, 1, \ldots, mN)$ are approximations of points $\mathbf{x}^{(m)}(t_k)$ of the periodic solution. In a neighborhood of $\mathbf{x}_k^{(m)*}$, with $\mathbf{x}_k^{(m)} = \mathbf{x}_k^{(m)*} + \Delta\mathbf{x}_k^{(m)}$, the linearized equation is given by

$$\Delta\mathbf{x}_k^{(m)} = DP_k \cdot \Delta\mathbf{x}_{k-1}^{(m)}$$
$$\text{with } \mathbf{g}_k(\mathbf{x}_{k-1}^{(m)*} + \Delta\mathbf{x}_{k-1}^{(m)}, \mathbf{x}_k^{(m)*} + \Delta\mathbf{x}_k^{(m)}, \mathbf{p}) = \mathbf{0}$$
$$(k = 1, 2, \ldots, mN). \tag{2.15}$$

The resultant Jacobian matrices of the periodic motion are

$$DP_{k(k-1)\cdots 1} = DP_k \cdot DP_{k-1} \cdot \ldots \cdot DP_1, \quad (k = 1, 2, \ldots, mN);$$

$$DP \equiv DP_{mN(mN-1)\cdots 1} = DP_{mN} \cdot DP_{mN-1} \cdot \ldots \cdot DP_1 \tag{2.16}$$

where

$$DP_k = \left[\frac{\partial \mathbf{x}_k^{(m)}}{\partial \mathbf{x}_{k-1}^{(m)}}\right]_{(\mathbf{x}_{k-1}^{(m)*},\mathbf{x}_k^{(m)*})} = -\left[\frac{\partial \mathbf{g}_k}{\partial \mathbf{x}_k^{(m)}}\right]^{-1}\left[\frac{\partial \mathbf{g}_k}{\partial \mathbf{x}_{k-1}^{(m)}}\right]\Bigg|_{(\mathbf{x}_{k-1}^{(m)*},\mathbf{x}_k^{(m)*})}. \tag{2.17}$$

The eigenvalues of $DP(\mathbf{x}_0^{(m)*})$ and $DP_{k(k-1)\cdots 1}$ for such a periodic motion are determined by

$$|DP_{k(k-1)\cdots 1} - \bar{\lambda}\mathbf{I}_{n\times n}| = 0, \quad (k = 1, 2, \ldots, mN);$$
$$|DP - \lambda\mathbf{I}_{n\times n}| = 0. \tag{2.18}$$

Thus, the stability and bifurcation of the periodic motion can be classified by the eigenvalues of $DP(\mathbf{x}_0^{(m)*})$ with

$$([n_1^m, n_1^o] : [n_2^m, n_2^o] : [n_3, \kappa_3] : [n_4, \kappa_4]|n_5 : n_6 : [n_7, l, \kappa_7]). \tag{2.19}$$

(i) *If the magnitudes of all eigenvalues of $DP^{(m)}$ are less than one (i.e., $|\lambda_i| < 1$, $i = 1, 2, \ldots, n$), the approximate period-m solution is stable.*

(ii) *If at least the magnitude of one eigenvalue of $DP^{(m)}$ is greater than one (i.e., $|\lambda_i| > 1$, $i \in \{1, 2, \ldots, n\}$), the approximate period-m solution is unstable.*

(iii) *The boundaries between stable and unstable period-m motions give bifurcation and stability conditions.*

To approximate a periodic motion in a nonlinear system, consider N-nodes of the periodic motion for $\|\mathbf{x}(t_k) - \mathbf{x}_k\| \leq \varepsilon_k$ $(\varepsilon_k \geq 0)$ and $\|\mathbf{f}(\mathbf{x}(t_k), t_k, \mathbf{p}) - \mathbf{f}(\mathbf{x}_k, t_k, \mathbf{p})\| \leq \delta_k$ $(\delta_k \geq 0)$, as shown in Fig. 2.1. $\delta = \max\{\delta_k\}_{k\in\{1,2,\ldots,N\}}$ and $\varepsilon = \max\{\varepsilon_k\}_{k\in\{1,2,\ldots,N\}}$ are the small prescribed positive quantities. The periodic motion can be approximated through a set of specific implicit mappings P_k with $\mathbf{g}_k(\mathbf{x}_{k-1}, \mathbf{x}_k, \mathbf{p}) = \mathbf{0}$ $(k = 1, 2, \ldots, N)$ with $\mathbf{x}_N = \mathbf{x}_0$. From the approximate mapping functions, the approximate trajectory nodes of the periodic motion are depicted by symbols connected by a solid curve. The exact solution of the periodic motion is given by a dashed curve. The discrete mapping P_k is developed from the differential equation. With controlling computational accuracy, the nodes of the periodic motion can be obtained with a good approximation.

If the period-doubling bifurcation of the periodic motion occurs, it will become a new periodic motion with a period of $T' = 2T$, which is called a period-2 motion. Due to the period-doubling, $2N$ nodes of the period-2 motion will be employed to describe the period-2 motion. Thus, consider a mapping structure of the period-2 motion with $2N$ implicit mappings.

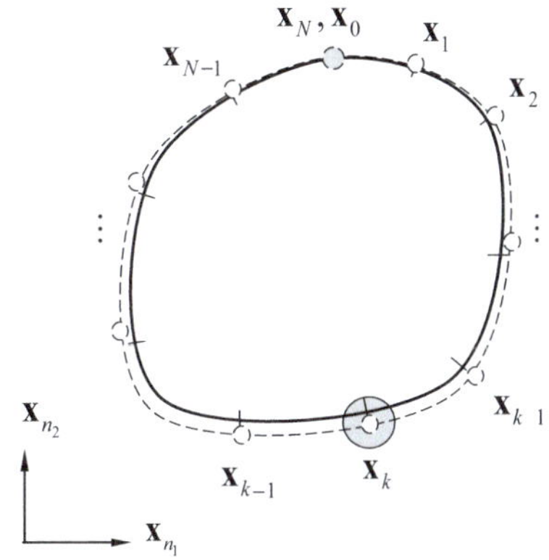

Fig. 2.1 Period-1 motion with N-node points. *Solid curve* numerical results, and dashed curve: expected exact results with N-nodes marked by symbols. The local shaded area is a small neighborhood of the exact solution at the kth node

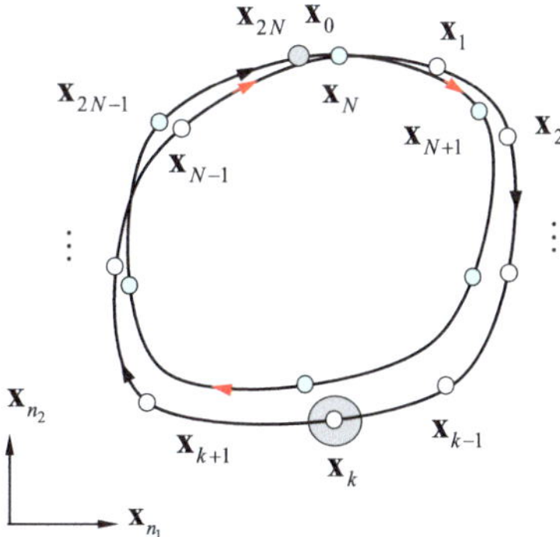

Fig. 2.2 Period-2 motion with $2N$-nodes. *Solid curve* numerical results. The symbols are node points on the periodic motion

$$P = P_{2N} \circ P_{2N-1} \circ \cdots \circ P_2 \circ P_1 : \mathbf{x}_0 \to \mathbf{x}_{2N};$$
$$\text{with } P_k : \mathbf{x}_{k-1} \to \mathbf{x}_k \quad (k = 1, 2, \ldots, 2N). \tag{2.20}$$

For $\mathbf{x}_{2N} = P\mathbf{x}_0$, there is a set of points $\mathbf{x}_k^*$ $(k = 0, 1, \ldots, 2N)$ computed by the following implicit vector functions

$$\mathbf{g}_k(\mathbf{x}_{k-1}^*, \mathbf{x}_k^*, \mathbf{p}) = \mathbf{0}, \quad (k = 1, 2, \ldots, 2N)$$
$$\mathbf{x}_0^* = \mathbf{x}_{2N}^*. \tag{2.21}$$

After period-doubling, the period-1 motion becomes period-2 motion. The nodes points increase to $2N$ points during two periods $(2T)$. The period-2 motion can be sketched in Fig. 2.2. The node points are determined through the discrete implicit mapping with a mathematical relation in Eq. (2.21). On the other hand,

$$T' = 2T = \frac{2(2\pi)}{\Omega} = \frac{2\pi}{\omega} \Rightarrow \omega = \frac{\Omega}{2}. \tag{2.22}$$

During the period of T', there is a periodic motion, which can be described by node points $\mathbf{x}_k$ ($k = 1, 2, \ldots, N'$). Since the period-1 motion is described by node points $\mathbf{x}_k$ ($k = 1, 2, \ldots, N$) during the period T, the period-2 motion can be described by $N' \geq 2N$ nodes. Thus the corresponding mapping P_k is defined as

$$P_k : \mathbf{x}_{k-1}^{(2)} \to \mathbf{x}_k^{(2)} \quad (k = 1, 2, \ldots, 2N) \tag{2.23}$$

and

$$\begin{aligned}
\mathbf{g}_k(\mathbf{x}_{k-1}^{(2)*}, \mathbf{x}_k^{(2)*}, \mathbf{p}) &= \mathbf{0}, \quad (k = 1, 2, \ldots, 2N) \\
\mathbf{x}_0^{(2)*} &= \mathbf{x}_{2N}^{(2)*}.
\end{aligned} \tag{2.24}$$

In general, for period $T' = mT$, there is a period-m motion which can be described by $N' \geq mN$. The corresponding mapping P_k is given by

$$P_k : \mathbf{x}_{k-1}^{(m)} \to \mathbf{x}_k^{(m)} \quad (k = 1, 2, \ldots, mN) \tag{2.25}$$

and

$$\begin{aligned}
\mathbf{g}_k^{(m)}(\mathbf{x}_{k-1}^{(m)*}, \mathbf{x}_k^{(m)*}, \mathbf{p}) &= \mathbf{0}, \quad (k = 1, 2, \ldots, mN) \\
\mathbf{x}_0^{(m)*} &= \mathbf{x}_{mN}^{(m)*}.
\end{aligned} \tag{2.26}$$

References

1. Luo ACJ (2015) Periodic flows to chaos based on implicit mappings of continuous nonlinear systems. International Journal of Bifurcation and chaos, **25**(3): Article No: 1550044 (62 pages)
2. Luo ACJ (2015) Discretization and Implicit Mapping Dynamics, Beijing/Dordrecht, HEP/Springer

A Nonlinear Spring-Pendulum

3

3.1 Physical Description

As in Guo and Luo [1], consider a periodically excited spring pendulum system as shown in Fig. 3.1, where k_1 and k_2 indicate the linear and nonlinear spring constants, respectively. The excitation force is given as $Q \cos \Omega t$ on the pendulum, where Q and Ω are excitation amplitude and frequency, respectively. The angular displacement of the pendulum is θ. The unstretched length of spring is L, and the linear displacement of the spring is x. The unit vectors on the horizontal and vertical directions are $\mathbf{i}$ and $\mathbf{j}$, respectively. From Fig. 3.1, the position vector of the pendulum can be defined as

$$\mathbf{r} = (L + x) \sin \theta \mathbf{i} - (L + x) \cos \theta \mathbf{j}. \tag{3.1}$$

The velocity vector is

$$\dot{\mathbf{r}} = [(L + x)\dot{\theta} \cos \theta + \dot{x} \sin \theta]\mathbf{i} + [(L + x)\dot{\theta} \sin \theta - \dot{x} \cos \theta]\mathbf{j}. \tag{3.2}$$

The kinetic energy is the calculated as

$$T = \tfrac{1}{2}m\dot{\mathbf{r}} \cdot \dot{\mathbf{r}} = \tfrac{1}{2}m(L + x)^2\dot{\theta}^2 + \tfrac{1}{2}m\dot{x}^2. \tag{3.3}$$

The potential energy of the spring is

$$V = \int k_1 x + k_2 x^3 dx = \tfrac{1}{2}k_1 x^2 + \tfrac{1}{4}k_2 x^4. \tag{3.3}$$

The Lagrangian of the system is

$$L_a = T - V = \tfrac{1}{2}m(L + x)^2\dot{\theta}^2 + \tfrac{1}{2}m\dot{x}^2 - \left(\tfrac{1}{2}k_1 x^2 + \tfrac{1}{4}k_2 x^4\right). \tag{3.5}$$

© The Author(s), under exclusive license to Springer Nature Switzerland AG 2023   15
Y. Guo and A. C. J. Luo, *Periodic Motions to Chaos in a Spring-Pendulum System*,
Synthesis Lectures on Mechanical Engineering,
https://doi.org/10.1007/978-3-031-17883-2_3

Fig. 3.1 Periodically excited
spring pendulum

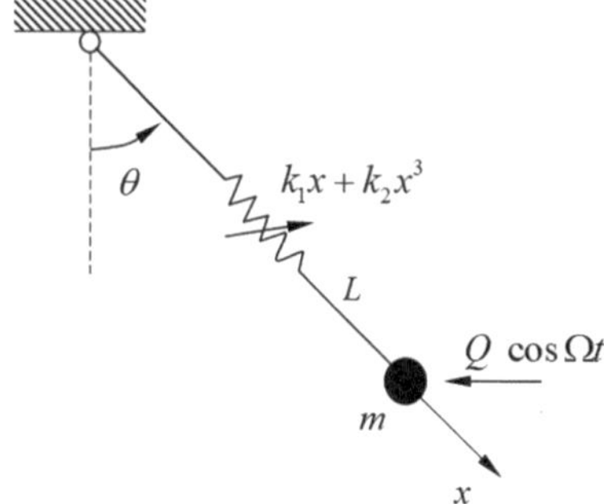

The damping force for the particle mass is

$$\mathbf{F}_d = -c[(L+x)\dot{\theta}\cos\theta + \dot{x}\sin\theta]\mathbf{i} - c[(L+x)\dot{\theta}\sin\theta - \dot{x}\cos\theta]\mathbf{j}, \qquad (3.6)$$

where c is the damping coefficient. The external force and gravity force for the particle
mass is

$$\mathbf{F}_c = -Q\cos\Omega t\,\mathbf{i} - mg\mathbf{j}. \qquad (3.7)$$

The resultant total force without spring for the particle is

$$\begin{aligned}
\mathbf{F} &= \mathbf{F}_c + \mathbf{F}_d \\
&= -[c(L+x)\dot{\theta}\cos\theta + c\dot{x}\sin\theta + Q\cos\Omega t]\mathbf{i} - [c(L+x)\dot{\theta}\sin\theta - c\dot{x}\cos\theta + mg]\mathbf{j}.
\end{aligned}$$
$$(3.8)$$

From the Lagrange equation,

$$\begin{aligned}
\frac{\partial}{\partial t}\left(\frac{\partial L_a}{\partial \dot{x}}\right) - \frac{\partial L_a}{\partial x} &= \mathbf{F}\cdot\frac{\partial \mathbf{r}}{\partial x}, \\
\frac{\partial}{\partial t}\left(\frac{\partial L_a}{\partial \dot{\theta}}\right) - \frac{\partial L_a}{\partial \theta} &= \mathbf{F}\cdot\frac{\partial \mathbf{r}}{\partial \theta}
\end{aligned} \qquad (3.9)$$

where

$$\frac{\partial L_a}{\partial x} = m(L+x)\dot{\theta}^2 - (k_1 x + k_2 x^3), \quad \frac{\partial L_a}{\partial \dot{x}} = m\dot{x}, \quad \frac{\partial}{\partial t}\left(\frac{\partial L_a}{\partial \dot{x}}\right) = m\ddot{x},$$

$$\frac{\partial \mathbf{r}}{\partial x} = \sin\theta\mathbf{i} - \cos\theta\mathbf{j}, \quad \mathbf{F}\cdot\frac{\partial \mathbf{r}}{\partial x} = mg\cos\theta - c\dot{x} - Q\cos\Omega t\sin\theta,$$

$$\frac{\partial L_a}{\partial \theta} = 0, \quad \frac{\partial L_a}{\partial \dot{\theta}} = m(L+x)^2\dot{\theta}, \quad \frac{\partial}{\partial t}\left(\frac{\partial L_a}{\partial \dot{\theta}}\right) = m(L+x)^2\ddot{\theta} + 2m\dot{x}(L+x)\dot{\theta},$$

$$\frac{\partial \mathbf{r}}{\partial \theta} = (L+x)\cos\theta\mathbf{i} + (L+x)\sin\theta\mathbf{j},$$

$$\mathbf{F} \cdot \frac{\partial \mathbf{r}}{\partial \theta} = -mg(L + x)\sin\theta - c\dot{\theta}(L + x)^2 - Q(L + x)\cos\Omega t \cos\theta. \tag{3.10}$$

From Eqs. (3.9) and (3.10), the equations of motions for such a spring pendulum system are obtained as

$$m\ddot{x} - m(L + x)\dot{\theta}^2 = mg\cos\theta - k_1 x - k_2 x^3 - c\dot{x} - Q\cos\Omega t \sin\theta,$$
$$m(L + x)^2\ddot{\theta} + 2m\dot{x}(L + x)\dot{\theta} = -mg(L + x)\sin\theta - c\dot{\theta}(L + x)^2$$
$$- Q(L + x)\cos\Omega t \cos\theta. \tag{3.11}$$

Such equations of motions in Eq. (3.11) are deformed as

$$\ddot{x} - (L + x)\dot{\theta}^2 = g\cos\theta - \alpha x - \beta x^3 - \delta\dot{x} - Q_0\cos\Omega t \sin\theta,$$
$$(L + x)^2\ddot{\theta} + 2\dot{x}(L + x)\dot{\theta} = -g(L + x)\sin\theta - \delta\dot{\theta}(L + x)^2 - Q_0(L + x)\cos\Omega t \cos\theta. \tag{3.12}$$

where $\delta = c/m$ is the damping coefficient, $\alpha = k_1/m$ and $\beta = k_2/m$ indicate the linear and nonlinear spring stiffness, respectively. $Q_0 = Q/m$ is excitation amplitude per unit mass. Such a system is expressed in state space as

$$\dot{x}_1 = y_1,$$
$$\dot{y}_1 = -\delta y_1 + (L + x_1)y_2^2 + g\cos x_2 - \alpha x_1 - \beta x_1^3 - Q_0\cos\Omega t \sin x_2,$$
$$\dot{x}_2 = y_2,$$
$$\dot{y}_2 = -\delta y_2 - \frac{2}{(L+x_1)}y_1 y_2 - \frac{1}{(L+x_1)}g\sin x_2 - \frac{Q_0}{(L+x_1)}\cos\Omega t \cos x_2, \tag{3.13}$$

where $x_1 = x$ and $x_2 = \theta$.

3.2   Discretization

As in [1–3] the above system can be discretized with a midpoint scheme for the time interval $t \in [t_k, t_{k+1}]$, forming an implicit map P_k ($k = 0, 1, 2, \ldots$) with

$$P_k : (x_{1(k-1)}, y_{1(k-1)}, x_{2(k-1)}, y_{2(k-1)}) \rightarrow (x_{1(k)}, y_{1(k)}, x_{2(k)}, y_{2(k)})$$
$$\Rightarrow (x_{1(k)}, y_{1(k)}, x_{2(k)}, y_{2(k)}) = P_k(x_{1(k-1)}, y_{1(k-1)}, x_{2(k-1)}, y_{2(k-1)}) \tag{3.14}$$

with the implicit relations

$$x_{1(k)} = x_{1(k-1)} + \tfrac{1}{2}h(y_{1(k-1)} + y_{1(k)}),$$
$$y_{1(k)} = y_{1(k-1)} + h\Big\{ -\tfrac{1}{2}\delta(y_{1(k-1)} + y_{1(k)}) + \tfrac{1}{4}\big[L + \tfrac{1}{2}(x_{1(k-1)} + x_{1(k)})\big](y_{2(k-1)} + y_{2(k)})^2$$
$$+ g\cos\tfrac{1}{2}(x_{2(k-1)} + x_{2(k)}) - \tfrac{1}{2}\alpha(x_{1(k-1)} + x_{1(k)}) - \tfrac{1}{8}\beta(x_{1(k-1)} + x_{1(k)})^3$$
$$- Q_0\cos\Omega\big(t_k + \tfrac{1}{2}h\big)\sin\tfrac{1}{2}(x_{2(k-1)} + x_{2(k)})\Big\},$$

$$x_{2(k)} = x_{2(k-1)} + \tfrac{1}{2}h(y_{2(k-1)} + y_{2(k)}),$$

$$\begin{aligned}
y_{2(k)} = y_{2(k-1)} + h\Big[&- \tfrac{1}{2}\delta(y_{2(k-1)} + y_{2(k)}) \\
&- [2L + (x_{1(k-1)} + x_{1(k)})]^{-1}(y_{1(k-1)} + y_{1(k)})(y_{2(k-1)} + y_{2(k)}) \\
&- \big[L + \tfrac{1}{2}(x_{1(k-1)} + x_{1(k)})\big]^{-1} g \sin \tfrac{1}{2}(x_{2(k-1)} + x_{2(k)}) \\
&- \big[L + \tfrac{1}{2}(x_{1(k-1)} + x_{1(k)})\big]^{-1} Q_0 \cos \Omega\big(t_k + \tfrac{h}{2}\big) \cos \tfrac{1}{2}(x_{2(k-1)} + x_{2(k)})\Big].
\end{aligned} \tag{3.15}$$

The accuracy of such discretization is $O(h^3)$ for each step. Thus, to ensure the computational error is less than 10^{-8}, one needs to maintain $h < 10^{-3}$.

Define

$$\mathbf{g}_k = (g_{k1}, g_{k2}, g_{k3}, g_{k4})^{\mathrm{T}},$$
$$\mathbf{x}_k = (x_{1(k)}, y_{1(k)}, x_{2(k)}, y_{2(k)})^{\mathrm{T}}. \tag{3.16}$$

The function of $\mathbf{g}_k = (g_{k1}, g_{k2}, g_{k3}, g_{k4})^{\mathrm{T}}$ for mapping P_k $(k = 0, 1, 2, \ldots)$ is given by

$$\begin{aligned}
g_{k1} = {}&x_{1(k)} - \big[x_{1(k-1)} + \tfrac{1}{2}h(y_{1(k-1)} + y_{1(k)})\big], \\
g_{k2} = {}&y_{1(k)} - \Big(y_{1(k-1)} + h\big\{ - \tfrac{1}{2}\delta(y_{1(k-1)} + y_{1(k)}) \\
&+ \tfrac{1}{4}\big[L + \tfrac{1}{2}(x_{1(k-1)} + x_{1(k)})\big](y_{2(k-1)} + y_{2(k)})^2 \\
&+ g \cos \tfrac{1}{2}(x_{2(k-1)} + x_{2(k)}) - \tfrac{1}{2}\alpha(x_{1(k-1)} + x_{1(k)}) - \tfrac{1}{8}\beta(x_{1(k-1)} + x_{1(k)})^3 \\
&- Q_0 \cos \Omega\big(t_k + \tfrac{1}{2}h\big) \sin \tfrac{1}{2}(x_{2(k-1)} + x_{2(k)})\big\}\Big), \\
g_{k3} = {}&x_{2(k)} - \big[x_{2(k-1)} + \tfrac{1}{2}h(y_{2(k-1)} + y_{2(k)})\big], \\
g_{k4} = {}&y_{2(k)} - \Big\{y_{2(k-1)} + h\big[- \tfrac{1}{2}\delta(y_{2(k-1)} + y_{2(k)}) \\
&- [2L + (x_{1(k-1)} + x_{1(k)})]^{-1}(y_{1(k-1)} + y_{1(k)})(y_{2(k-1)} + y_{2(k)}) \\
&- \big[L + \tfrac{1}{2}(x_{1(k-1)} + x_{1(k)})\big]^{-1} g \sin \tfrac{1}{2}(x_{2(k-1)} + x_{2(k)}) \\
&- \big[L + \tfrac{1}{2}(x_{1(k-1)} + x_{1(k)})\big]^{-1} Q_0 \cos \Omega\big(t_k + \tfrac{h}{2}\big) \cos \tfrac{1}{2}(x_{2(k-1)} + x_{2(k)})\big]\Big\}.
\end{aligned} \tag{3.17}$$

Thus, Eq. (3.15) becomes

$$\mathbf{g}_k(\mathbf{x}_{k-1}, \mathbf{x}_k, \mathbf{p}) = \mathbf{0} \tag{3.18}$$

or the scalar form is

$$\begin{aligned}
g_{k1}(\mathbf{x}_{k-1}, \mathbf{x}_k, \mathbf{p}) &= 0, \\
g_{k2}(\mathbf{x}_{k-1}, \mathbf{x}_k, \mathbf{p}) &= 0, \\
g_{k3}(\mathbf{x}_{k-1}, \mathbf{x}_k, \mathbf{p}) &= 0, \\
g_{k4}(\mathbf{x}_{k-1}, \mathbf{x}_k, \mathbf{p}) &= 0.
\end{aligned} \tag{3.19}$$

References

1. Guo Y, Luo ACJ (2022) Period-3 motions to chaos in a periodically forced nonlinear-spring pendulum. AIP Chao, **32**, 103129
2. Luo ACJ, Yuan YG (2020) Bifurcation trees of period-1 to period-2 motions in a periodically excited nonlinear spring pendulum. Journal of Vibration Testing and System Dynamics, **4**(3): 201–248.
3. Guo Y (2022) Bifurcations and harmonic responses of period-1 motions in a periodically excited spring pendulum. Journal of Vibration Testing and System Dynamics, **6**(3): 297–315.

Formulation for Periodic Motions

4

4.1 Period-1 Motions

As in [1–3], a period-1 motion in the periodically excited spring pendulum can be represented by a discrete mapping structure of N mapping actions:

$$P = \underbrace{P_N \circ P_{N-1} \circ \cdots \circ P_2 \circ P_1}_{N\text{-actions}} : \mathbf{x}_0 \to \mathbf{x}_N \qquad (4.1)$$

with

$$P_k : \mathbf{x}_{k-1} \to \mathbf{x}_k \Rightarrow \mathbf{x}_k = P_k \mathbf{x}_{k-1} \quad (k = 1, 2, \ldots, N) \qquad (4.2)$$

From Eq. (3.15), the corresponding algebraic equations are obtained by

$$
\begin{aligned}
x_{1(k)} ={}& x_{1(k-1)} + \tfrac{1}{2}h\big(y_{1(k-1)} + y_{1(k)}\big), \\
y_{1(k)} ={}& y_{1(k-1)} + h\Big[-\tfrac{1}{2}\delta\big(y_{1(k-1)} + y_{1(k)}\big) + \tfrac{1}{4}\big(L + \tfrac{1}{2}(x_{1(k-1)} + x_{1(k)})\big)\big(y_{2(k-1)} + y_{2(k)}\big)^2 \\
&+ g\cos\tfrac{1}{2}(x_{2(k-1)} + x_{2(k)}) - \tfrac{1}{2}\alpha(x_{1(k-1)} + x_{1(k)}) - \tfrac{1}{8}\beta(x_{1(k-1)} + x_{1(k)})^3 \\
&- Q_0 \cos\Omega\big(t_{k-1} + \tfrac{1}{2}h\big) \sin\tfrac{1}{2}(x_{2(k-1)} + x_{2(k)})\Big], \\
x_{2(k)} ={}& x_{2(k-1)} + \tfrac{1}{2}h\big(y_{2(k-1)} + y_{2(k)}\big), \\
y_{2(k)} ={}& y_{2(k-1)} + h\Big[-\tfrac{1}{2}\delta\big(y_{2(k-1)} + y_{2(k)}\big) - \big[2L + (x_{1(k-1)} + x_{1(k)})\big]^{-1} \\
&\times (y_{1(k-1)} + y_{1(k)})(y_{2(k)} + y_{2(k)}) \\
&- \big[L + \tfrac{1}{2}(x_{1(k-1)} + x_{1(k)})\big]^{-1} g \sin\tfrac{1}{2}(x_{2(k-1)} + x_{2(k)}) \\
&- \big[L + \tfrac{1}{2}(x_{1(k-1)} + x_{1(k)})\big]^{-1} Q_0 \cos\Omega\,(t_{k-1} + \tfrac{1}{2}h)\cos\tfrac{1}{2}(x_{2(k-1)} + x_{2(k)})\Big]; \\
& \text{for } P_k \quad (k = 1, 2, \ldots, N)
\end{aligned}
\qquad (4.3)
$$

© The Author(s), under exclusive license to Springer Nature Switzerland AG 2023

Y. Guo and A. C. J. Luo, *Periodic Motions to Chaos in a Spring-Pendulum System*, Synthesis Lectures on Mechanical Engineering, https://doi.org/10.1007/978-3-031-17883-2_4

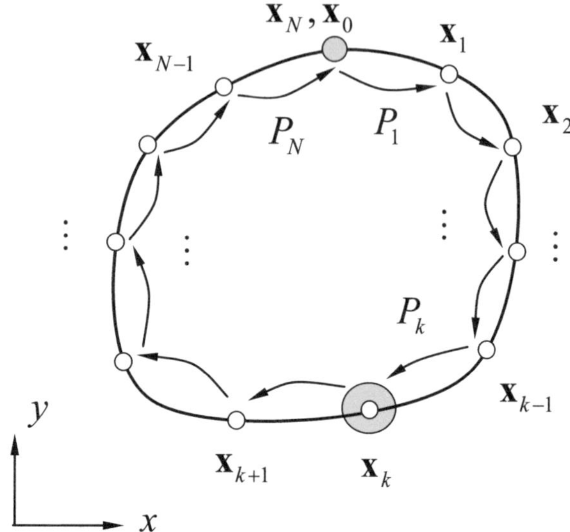

Fig. 4.1 Mapping structure of period-1 motions with N-nodes

The corresponding periodicity conditions of $\mathbf{x}_0 = \mathbf{x}_N$ is given as

$$x_{1(N)} = x_{1(0)}, \quad y_{1(N)} = y_{1(0)},$$
$$x_{2(N)} = x_{2(0)} + 2n\pi \quad (n = 0, 1, 2, \ldots),$$
$$y_{2(N)} = y_{2(0)}. \tag{4.4}$$

With vector $\mathbf{x} = (x_1, y_1, x_2, y_2)^{\mathrm{T}}$, the mapping structure for period-1 motions are illustrated in Fig. 2.2 through N-nodes.

From Eqs. (4.3) and (4.4), all the nodes on the discretized orbit can be determined by $4(N + 1)$ equations. Once such node points $\mathbf{x}_k^*$ $(k = 1, 2, \ldots, N)$ of the period-m motion are obtained, the stability of period-m motion can be discussed through the corresponding Jacobian matrix. For a small perturbation in vicinity of $\mathbf{x}_k^*$, $\mathbf{x}_k = \mathbf{x}_k^* + \Delta\mathbf{x}_k$, $(k = 0, 1, 2, \ldots, N)$,

$$\Delta\mathbf{x}_N = DP\Delta\mathbf{x}_0 = \underbrace{DP_N \cdot DP_{N-1} \cdot \ldots \cdot DP_2 \cdot DP_1}_{N\text{-muplication}} \Delta\mathbf{x}_0. \tag{4.5}$$

with

$$\Delta\mathbf{x}_k = DP_k\Delta\mathbf{x}_{k-1} \equiv \left[\frac{\partial\mathbf{x}_k}{\partial\mathbf{x}_{k-1}}\right]_{(\mathbf{x}_{k-1}^*, \mathbf{x}_k^*)} \Delta\mathbf{x}_{k-1},$$
$$(k = 1, 2, \ldots, N) \tag{4.6}$$

where

$$DP_k = \left[\frac{\partial \mathbf{x}_k}{\partial \mathbf{x}_{k-1}}\right]_{(\mathbf{x}_{k-1}^*, \mathbf{x}_k^*)} = \begin{bmatrix} \dfrac{\partial x_{1(k)}}{\partial x_{1(k-1)}} & \dfrac{\partial x_{1(k)}}{\partial y_{1(k-1)}} & \dfrac{\partial x_{1(k)}}{\partial x_{2(k-1)}} & \dfrac{\partial x_{1(k)}}{\partial y_{2(k-1)}} \\[2mm] \dfrac{\partial y_{1(k)}}{\partial x_{1(k-1)}} & \dfrac{\partial y_{1(k)}}{\partial y_{1(k-1)}} & \dfrac{\partial y_{1(k)}}{\partial x_{2(k-1)}} & \dfrac{\partial y_{1(k)}}{\partial y_{2(k-1)}} \\[2mm] \dfrac{\partial x_{2(k)}}{\partial x_{1(k-1)}} & \dfrac{\partial x_{2(k)}}{\partial y_{1(k)}} & \dfrac{\partial x_{2(k)}}{\partial x_{2(k)}} & \dfrac{\partial x_{2(k)}}{\partial y_{2(k-1)}} \\[2mm] \dfrac{\partial y_{2(k)}}{\partial x_{1(k-1)}} & \dfrac{\partial y_{2(k)}}{\partial y_{1(k)}} & \dfrac{\partial y_{2(k)}}{\partial x_{2(k-1)}} & \dfrac{\partial y_{2(k)}}{\partial y_{2(k-1)}} \end{bmatrix}_{(\mathbf{x}_{k-1}^*, \mathbf{x}_k^*)}$$

$$\text{for } k = 1, 2, \ldots, N. \tag{4.7}$$

and

$$DP_k = \frac{\partial \mathbf{x}_k}{\partial \mathbf{x}_{k-1}}\Big|_{(\mathbf{x}_{k-1}^*, \mathbf{x}_k^*)} = -\left(\frac{\partial \mathbf{g}_k}{\partial \mathbf{x}_k}\right)^{-1} \frac{\partial \mathbf{g}_k}{\partial \mathbf{x}_{k-1}}\Big|_{(\mathbf{x}_{k-1}^*, \mathbf{x}_k^*)}, \tag{4.8}$$

the corresponding matrices $\partial \mathbf{g}_k^{(m)}/\partial \mathbf{x}_k^{(m)}$ and $\partial \mathbf{g}_k^{(m)}/\partial \mathbf{x}_{k-1}^{(m)}$ are given by

$$\frac{\partial \mathbf{g}_k}{\partial \mathbf{x}_k} = \begin{bmatrix} \dfrac{\partial g_{k1}}{\partial x_{1(k)}} & \dfrac{\partial g_{k1}}{\partial y_{1(k)}} & \dfrac{\partial g_{k1}}{\partial x_{2(k)}} & \dfrac{\partial g_{k1}}{\partial y_{2(k)}} \\[2mm] \dfrac{\partial g_{k2}}{\partial x_{1(k)}} & \dfrac{\partial g_{k2}}{\partial y_{1(k)}} & \dfrac{\partial g_{k2}}{\partial x_{2(k)}} & \dfrac{\partial g_{k2}}{\partial y_{2(k)}} \\[2mm] \dfrac{\partial g_{k3}}{\partial x_{1(k)}} & \dfrac{\partial g_{k3}}{\partial y_{1(k)}} & \dfrac{\partial g_{k3}}{\partial x_{2(k)}} & \dfrac{\partial g_{k3}}{\partial y_{2(k)}} \\[2mm] \dfrac{\partial g_{k4}}{\partial x_{1(k)}} & \dfrac{\partial g_{k4}}{\partial y_{1(k)}} & \dfrac{\partial g_{k4}}{\partial x_{2(k)}} & \dfrac{\partial g_{k4}}{\partial y_{2(k)}} \end{bmatrix} \tag{4.9}$$

where

$$\frac{\partial g_{k1}}{\partial x_{1(k)}} = 1, \quad \frac{\partial g_{k3}}{\partial x_{1(k)}} = 0,$$

$$\frac{\partial g_{k2}}{\partial x_{1(k)}} = -\tfrac{1}{8}h(y_{2(k-1)} + y_{2(k)})^2 + \frac{1}{2m_1}hk_1 + \frac{3}{8m_1}hk_2(x_{1(k-1)} + x_{1(k)})^2,$$

$$\frac{\partial g_{k4}}{\partial x_{1(k)}} = \left[L + \tfrac{1}{2}(x_{1(k-1)} + x_{1(k)})\right]^{-2}\Big\{ -\frac{1}{2m_1}hQ_0 \cos\Omega\big(t_{k-1} + \tfrac{1}{2}h\big)$$

$$\times \cos\left[\tfrac{1}{2}(x_{2(k-1)} + x_{2(k)})\right] - \frac{h}{4}(y_{1(k-1)} + y_{1(k)})(y_{2(k-1)} + y_{2(k)})$$

$$- \tfrac{1}{2}hg \sin\left[\tfrac{1}{2}(x_{2(k-1)} + x_{2(k)})\right]\Big\}, \tag{4.10}$$

$$\frac{\partial g_{k1}}{\partial y_{1(k)}} = -\tfrac{1}{2}h,$$

$$\frac{\partial g_{k2}}{\partial y_{1(k)}} = 1 + \frac{1}{2m_1}hc,$$

$$\frac{\partial g_{k3}}{\partial y_{1(k)}} = 0,$$

$$\frac{\partial g_{k4}}{\partial y_{1(k)}} = \tfrac{1}{2}h\left[L + \tfrac{1}{2}(x_{1(k-1)} + x_{1(k)})\right]^{-1}(y_{2(k-1)} + y_{2(k)}). \tag{4.11}$$

$$\frac{\partial g_{k1}}{\partial x_{2(k)}} = 0,$$

$$\frac{\partial g_{k2}}{\partial x_{2(k)}} = \tfrac{1}{2}hg\sin\left[\tfrac{1}{2}(x_{2(k-1)} + x_{2(k)})\right] + \frac{1}{2m_1}hQ_0\cos\Omega\left(t_{k-1} + \tfrac{1}{2}h\right)$$

$$\times \cos\left[\tfrac{1}{2}(x_{2(k-1)} + x_{2(k)})\right],$$

$$\frac{\partial g_{k3}}{\partial x_{2(k)}} = 1,$$

$$\frac{\partial g_{k4}}{\partial x_{2(k)}} = \left[L + \tfrac{1}{2}(x_{1(k-1)} + x_{1(k)})\right]^{-1}\left\{\tfrac{1}{2}hg\cos\left[\tfrac{1}{2}(x_{2(k-1)} + x_{2(k)})\right]\right.$$

$$\left. - \frac{1}{2m_1}hQ_0\cos\Omega(t_{k-1} + \tfrac{1}{2}h)\sin\left[\tfrac{1}{2}(x_{2(k-1)} + x_{2(k)})\right]\right\}; \tag{4.12}$$

$$\frac{\partial g_{k1}}{\partial y_{2k}} = 0,$$

$$\frac{\partial g_{k2}}{\partial y_{2k}} = -\tfrac{1}{2}\left[L + \tfrac{1}{2}(x_{1(k-1)} + x_{1k})\right](y_{1(k-1)} + y_{1k}),$$

$$\frac{\partial g_{k2}}{\partial y_{2k}} = -\tfrac{1}{2}h,$$

$$\frac{\partial g_{k2}}{\partial y_{2k}} = 1 + \frac{1}{2m_1}hc + \tfrac{1}{2}h\left[L + \tfrac{1}{2}(x_{1(k-1)} + x_{1k})\right]^{-1}(y_{1(k-1)} + y_{1k}); \tag{4.13}$$

and

$$\frac{\partial \mathbf{g}_k}{\partial \mathbf{x}_{k-1}} = \begin{bmatrix} \frac{\partial g_{k1}}{\partial x_{1(k-1)}} & \frac{\partial g_{k1}}{\partial y_{1(k-1)}} & \frac{\partial g_{k1}}{\partial x_{2(k-1)}} & \frac{\partial g_{k1}}{\partial y_{2(k-1)}} \\ \frac{\partial g_{k2}}{\partial x_{1(k-1)}} & \frac{\partial g_{k2}}{\partial y_{1(k-1)}} & \frac{\partial g_{k2}}{\partial x_{2(k-1)}} & \frac{\partial g_{k2}}{\partial y_{2(k-1)}} \\ \frac{\partial g_{k3}}{\partial x_{1(k-1)}} & \frac{\partial g_{k3}}{\partial y_{1(k-1)}} & \frac{\partial g_{k3}}{\partial x_{2(k-1)}} & \frac{\partial g_{k3}}{\partial y_{2(k-1)}} \\ \frac{\partial g_{k4}}{\partial x_{1(k-1)}} & \frac{\partial g_{k4}}{\partial y_{1(k-1)}} & \frac{\partial g_{k4}}{\partial x_{2(k-1)}} & \frac{\partial g_{k4}}{\partial y_{2(k-1)}} \end{bmatrix} \tag{4.14}$$

where

$$\frac{\partial g_{k1}}{\partial x_{1(k-1)}} = -1, \quad \frac{\partial g_{k3}}{\partial x_{1(k-1)}} = 0,$$

$$\frac{\partial g_{k2}}{\partial x_{1(k-1)}} = -\tfrac{1}{8}h(y_{2(k-1)} + y_{2(k)})^2 + \frac{1}{2m_1}hk_1 + \frac{3}{8m_1}hk_2(x_{1(k-1)} + x_{1(k)})^2,$$

$$\frac{\partial g_{k4}}{\partial x_{1(k-1)}} = \left[L + \tfrac{1}{2}(x_{1(k-1)} + x_{1(k)})\right]^{-2}\left\{-\frac{1}{2m_1}hQ_0\cos\Omega\left(t_{k-1} + \tfrac{1}{2}h\right)\right.$$

$$\times \cos\left[\tfrac{1}{2}(x_{2(k-1)} + x_{2(k)})\right] - \tfrac{h}{4}(y_{1(k-1)} + y_{1(k)})(y_{2(k-1)} + y_{2(k)})$$

$$\left. - \tfrac{1}{2}hg\sin\left[\tfrac{1}{2}(x_{2(k-1)} + x_{2(k)})\right]\right\}. \tag{4.15}$$

$$\frac{\partial g_{k1}}{\partial y_{1(k-1)}} = -\tfrac{1}{2}h,$$

$$\frac{\partial g_{k2}}{\partial y_{1(k-1)}} = -1 + \frac{1}{2m_1}hc,$$

$$\frac{\partial g_{k3}}{\partial y_{1(k-1)}} = 0,$$

$$\frac{\partial g_{k4}}{\partial y_{1(k-1)}} = \frac{h}{2}\big[L + \tfrac{1}{2}(x_{1(k-1)} + x_{1(k)})\big]^{-1}(y_{2(k-1)} + y_{2(k)}); \tag{4.16}$$

$$\frac{\partial g_{k1}}{\partial x_{2(k-1)}} = 0,$$

$$\frac{\partial g_{k2}}{\partial x_{2(k-1)}} = \tfrac{1}{2}gh \sin\big[\tfrac{1}{2}(x_{2(k-1)} + x_{2(k)})\big] + \tfrac{1}{2m_1}hQ_0 \cos\Omega\big(t_{k-1} + \tfrac{1}{2}h\big)$$

$$\times \cos\big[\tfrac{1}{2}(x_{2(k-1)} + x_{2(k)})\big],$$

$$\frac{\partial g_{k3}}{\partial x_{2(k-1)}} = -1,$$

$$\frac{\partial g_{k4}}{\partial x_{2(k-1)}} = \tfrac{1}{2}gh\big[L + \tfrac{1}{2}(x_{1(k-1)} + x_{1(k)})\big]^{-1}\Big\{ \cos\big[\tfrac{1}{2}(x_{2(k-1)} + x_{2(k)})\big]$$

$$- \tfrac{1}{m_1}Q_0 \cos\Omega\big(t_{k-1} + \tfrac{1}{2}h\big) \sin\big[\tfrac{1}{2}(x_{2(k-1)} + x_{2(k)})\big]\Big\}; \tag{4.17}$$

$$\frac{\partial g_{k1}}{\partial y_{2(k-1)}} = 0,$$

$$\frac{\partial g_{k2}}{\partial y_{2(k-1)}} = -\tfrac{1}{2}h\big[L + \tfrac{1}{2}(x_{1(k-1)} + x_{1(k)})\big](y_{2(k-1)} + y_{2(k)}),$$

$$\frac{\partial g_{k3}}{\partial y_{2(k-1)}} = -\tfrac{1}{2}h,$$

$$\frac{\partial g_{k4}}{\partial y_{2(k-1)}} = -1 + \tfrac{1}{2m_1}hc + \tfrac{1}{2}h\big[L + \tfrac{1}{2}(x_{1(k-1)} + x_{1(k)})\big]^{-1} \times (y_{1(k-1)} + y_{1(k)}). \tag{4.18}$$

The stability and bifurcation conditions of such period-m motion are measured through the eigenvalues, which are computed by

$$|DP - \lambda \mathbf{I}_{4\times4}| = 0 \tag{4.19}$$

where

$$DP = \prod_{k=N}^{1} \left[\frac{\partial \mathbf{x}_k}{\partial \mathbf{x}_{k-1}}\right]_{(\mathbf{x}_k^*, \mathbf{x}_{k-1}^*)}. \tag{4.20}$$

The corresponding stability conditions are given as:

(i) If all $|\lambda_i| < 1$ for $(i = 1, 2, 3, 4)$, the periodic motion is stable.
(ii) If one of $|\lambda_i| > 1$ for $(i \in \{1, 2, 3, 4\})$, the periodic motion is unstable.

The bifurcation points are identified as:

(i)   If one of $\lambda_i = -1$ for ($i \in \{1, 2, 3, 4\}$), the period-doubling bifurcation occurs.
(ii) If one of $\lambda_i = 1$ for ($i \in \{1, 2, 3, 4\}$), the saddle-node bifurcation occurs.
(iii) If $|\lambda_{i,j}| = 1$ with $\lambda_{i,j} = \alpha \pm i\beta$, and $|\lambda_l| < 1$ for (i, j, $l \in \{1, 2, 3, 4\}$), the Neimark bifurcation occurs.

4.2   Period-m Motions

In general, a period-m periodic motion in the periodically excited spring pendulum can be represented by a discrete mapping structure of mN mapping actions:

$$P = \underbrace{P_{mN} \circ P_{mN-1} \circ \cdots \circ P_2 \circ P_1}_{mN\text{-actions}} : (x_{1(0)}^{(m)}, y_{1(0)}^{(m)}, x_{2(0)}^{(m)}, y_{2(0)}^{(m)})$$

$$\rightarrow (x_{1(mN)}^{(m)}, y_{1(mN)}^{(m)}, x_{2(mN)}^{(m)}, y_{2(mN)}^{(m)}) \tag{4.21}$$

with

$$P_k : (x_{1(k-1)}^{(m)}, y_{1(k-1)}^{(m)}, x_{2(k-1)}^{(m)}, y_{2(k-1)}^{(m)}) \rightarrow (x_{1(k)}^{(m)}, y_{1(k)}^{(m)}, x_{2(k)}^{(m)}, y_{2(k)}^{(m)})$$

$$\Rightarrow (x_{1(k)}^{(m)}, y_{1(k)}^{(m)}, x_{2(k)}^{(m)}, y_{2(k)}^{(m)}) = P_k(x_{1(k-1)}^{(m)}, y_{1(k-1)}^{(m)}, x_{2(k-1)}^{(m)}, y_{2(k-1)}^{(m)})$$

$$(k = 1, 2, \ldots, mN) \tag{4.22}$$

From Eq. (3.15), the corresponding algebraic equations are obtained by

$$x_{1(k)}^{(m)} = x_{1(k-1)}^{(m)} + \tfrac{1}{2}h(y_{1(k-1)}^{(m)} + y_{1(k)}^{(m)}),$$

$$y_{1(k)}^{(m)} = y_{1(k-1)}^{(m)} + h\big[-\tfrac{1}{2}\delta(y_{1(k-1)}^{(m)} + y_{1(k)}^{(m)}) + \tfrac{1}{4}\big(L + \tfrac{1}{2}(x_{1(k-1)}^{(m)} + x_{1(k)}^{(m)})\big)(y_{2(k-1)}^{(m)} + y_{2(k)}^{(m)})^2$$

$$+ g\cos\tfrac{1}{2}(x_{2(k-1)}^{(m)} + x_{2(k)}^{(m)}) - \tfrac{1}{2}\alpha(x_{1(k-1)}^{(m)} + x_{1(k)}^{(m)}) - \tfrac{1}{8}\beta(x_{1(k-1)}^{(m)} + x_{1(k)}^{(m)})^3$$

$$- Q_0\cos\Omega\big(t_{k-1} + \tfrac{1}{2}h\big)\sin\tfrac{1}{2}(x_{2(k-1)}^{(m)} + x_{2(k)}^{(m)})\big],$$

$$x_{2(k)}^{(m)} = x_{2(k-1)}^{(m)} + \tfrac{1}{2}h(y_{2(k-1)}^{(m)} + y_{2(k)}^{(m)}),$$

$$y_{2(k)}^{(m)} = y_{2(k-1)}^{(m)} + h\big[-\tfrac{1}{2}\delta(y_{2(k-1)}^{(m)} + y_{2(k)}^{(m)}) - [2L + (x_{1(k-1)}^{(m)} + x_{1(k)}^{(m)})]^{-1}$$

$$\times (y_{1(k-1)}^{(m)} + y_{1(k)}^{(m)})(y_{2(k-1)}^{(m)} + y_{2(k)}^{(m)})$$

$$- \big[L + \tfrac{1}{2}(x_{1(k-1)}^{(m)} + x_{1(k)}^{(m)})\big]^{-1} g\sin\tfrac{1}{2}(x_{2(k-1)}^{(m)} + x_{2(k)}^{(m)})$$

$$- \big[L + \tfrac{1}{2}(x_{1(k-1)}^{(m)} + x_{1(k)}^{(m)})\big]^{-1} Q_0\cos\Omega\big(t_{k-1} + \tfrac{1}{2}h\big)\cos\tfrac{1}{2}(x_{2(k-1)}^{(m)} + x_{2(k)}^{(m)})\big];$$

$$\text{for } P_k \quad (k = 1, 2, \ldots mN) \tag{4.23}$$

The corresponding periodicity conditions is given as

$$(x_{1(mN)}^{(m)}, y_{1(mN)}^{(m)}, x_{2(mN)}^{(m)}, y_{2(mN)}^{(m)}) = (x_{1(0)}^{(m)}, y_{1(0)}^{(m)}, x_{2(0)}^{(m)} + 2n\pi, y_{2(0)}^{(m)}) \tag{4.24}$$

for $(n = 0, 1, 2, \ldots)$. With vector $\mathbf{x} = (x_1, y_1, x_2, y_2)^{\mathrm{T}}$, the mapping structure for period-2 motions is illustrated in Fig. 2.2 through $2N$-nodes to illustrate period-m motions.

From Eqs. (4.23) and (4.24), all the nodes on the discretized orbit can be determined by $4(mN + 1)$ equations. Once such node points $\mathbf{x}_k^{(m)*}$ $(k = 1, 2, \ldots, mN)$ of the period-m motion are obtained, the stability of period-m motion can be discussed through the corresponding Jacobian matrix. For a small perturbation in vicinity of $\mathbf{x}_k^{(m)*}$, $\mathbf{x}_k^{(m)} = \mathbf{x}_k^{(m)*} + \Delta\mathbf{x}_k^{(m)}$, $(k = 0, 1, 2, \ldots, mN)$,

$$\Delta\mathbf{x}_{mN} = DP\Delta\mathbf{x}_0^{(m)} = \underbrace{DP_{mN} \cdot DP_{mN-1} \cdot \ldots \cdot DP_2 \cdot DP_1}_{mN\text{-muplication}} \Delta\mathbf{x}_0^{(m)}. \tag{4.25}$$

with

$$\Delta\mathbf{x}_k^{(m)} = DP_k \Delta\mathbf{x}_{k-1}^{(m)} \equiv \left[\frac{\partial\mathbf{x}_k^{(m)}}{\partial\mathbf{x}_{k-1}^{(m)}}\right]_{(\mathbf{x}_k^{(m)*},\mathbf{x}_{k-1}^{(m)*})} \Delta\mathbf{x}_{k-1}^{(m)},$$

$$(k = 1, 2, \ldots, mN) \tag{4.26}$$

where

$$DP_k = \left[\frac{\partial\mathbf{x}_k^{(m)}}{\partial\mathbf{x}_{k-1}^{(m)}}\right]_{(\mathbf{x}_{k-1}^{(m)*},\mathbf{x}_k^{(m)*})} = \begin{bmatrix} \dfrac{\partial x_{1(k)}^{(m)}}{\partial x_{1(k-1)}^{(m)}} & \dfrac{\partial x_{1(k)}^{(m)}}{\partial y_{1(k-1)}^{(m)}} & \dfrac{\partial x_{1(k)}^{(m)}}{\partial x_{2(k-1)}^{(m)}} & \dfrac{\partial x_{1(k)}^{(m)}}{\partial y_{2(k-1)}^{(m)}} \\[2ex] \dfrac{\partial y_{1(k)}^{(m)}}{\partial x_{1(k-1)}^{(m)}} & \dfrac{\partial y_{1(k)}^{(m)}}{\partial y_{1(k-1)}^{(m)}} & \dfrac{\partial y_{1(k)}^{(m)}}{\partial x_{2(k-1)}^{(m)}} & \dfrac{\partial y_{1(k)}^{(m)}}{\partial y_{2(k-1)}^{(m)}} \\[2ex] \dfrac{\partial x_{2(k)}^{(m)}}{\partial x_{1(k-1)}^{(m)}} & \dfrac{\partial x_{2(k)}^{(m)}}{\partial y_{1(k-1)}^{(m)}} & \dfrac{\partial x_{2(k)}^{(m)}}{\partial x_{2(k-1)}^{(m)}} & \dfrac{\partial x_{2(k)}^{(m)}}{\partial y_{2(k-1)}^{(m)}} \\[2ex] \dfrac{\partial y_{2(k)}^{(m)}}{\partial x_{1(k-1)}^{(m)}} & \dfrac{\partial y_{2(k)}^{(m)}}{\partial y_{1(k-1)}^{(m)}} & \dfrac{\partial y_{2(k)}^{(m)}}{\partial x_{2(k-1)}^{(m)}} & \dfrac{\partial y_{2(k)}^{(m)}}{\partial y_{2(k-1)}^{(m)}} \end{bmatrix}_{(\mathbf{x}_{k-1}^{(m)*},\mathbf{x}_k^{(m)*})}$$

$$\text{for } k = 1, 2, \ldots, mN. \tag{4.27}$$

Similarly, the variational equation of Eq. (3.18) gives

$$\Delta\mathbf{x}_k^{(m)} = DP_k \Delta\mathbf{x}_{k-1}^{(m)} \tag{4.28}$$

and for $k = 1, 2, \ldots, mN$

$$DP_k = \frac{\partial\mathbf{x}_k^{(m)}}{\partial\mathbf{x}_{k-1}^{(m)}}\Big|_{(\mathbf{x}_{k-1}^{(m)*},\mathbf{x}_k^{(m)*})} = -\left(\frac{\partial\mathbf{g}_k^{(m)}}{\partial\mathbf{x}_k^{(m)}}\right)^{-1} \frac{\partial\mathbf{g}_k^{(m)}}{\partial\mathbf{x}_{k-1}^{(m)}}\Big|_{(\mathbf{x}_{k-1}^{(m)*},\mathbf{x}_k^{(m)*})} \tag{4.29}$$

where $\partial\mathbf{g}_k^{(m)}/\partial\mathbf{x}_k^{(m)}$ and $\partial\mathbf{g}_k^{(m)}/\partial\mathbf{x}_{k-1}^{(m)}$ are similar to Eqs. (4.9)–(4.18). The resultant Jacobian matrix of the period-m motion is

$$DP = DP_{mN(N-1)\cdots 1} = DP_{mN} \cdot DP_{mN-1} \cdot \ldots \cdot DP_1. \tag{4.30}$$

The eigenvalues of DP for the period-m motion are computed by

$$|DP - \lambda \mathbf{I}_{4\times 4}| = 0 \tag{4.31}$$

where

$$DP = \prod_{k=mN}^{1} \left[\frac{\partial \mathbf{x}_k^{(m)}}{\partial \mathbf{x}_{k-1}^{(m)}} \right]_{(\mathbf{x}_k^{(m)*}, \mathbf{x}_{k-1}^{(m)*})}. \tag{4.32}$$

The corresponding stability and bifurcation conditions are identical to the period-1 motion.

4.3   Finite-Fourier Series

From the node points of period-m motions $\mathbf{x}_k^{(m)} = (x_{1(k)}^{(m)}, y_{1(k)}^{(m)}, x_{2(k)}^{(m)}, y_{2(k)}^{(m)})^{\mathrm{T}}$ ($k = 0, 1, 2, \ldots, mN$) in the parametrically excited pendulum, as in Luo [4], the period-m motions can then be approximately expressed by the Fourier series, i.e.,

$$\mathbf{x}^{(m)}(t) \approx \mathbf{a}_0^{(m)} + \sum_{j=1}^{M} \mathbf{b}_{j/m} \cos\left(\frac{j}{m}\Omega t\right) + \mathbf{c}_{j/m} \sin\left(\frac{j}{m}\Omega t\right). \tag{4.33}$$

The $(2M+1)$ unknown vector coefficients of $\mathbf{a}_0^{(m)}$, $\mathbf{b}_{j/m}$, $\mathbf{c}_{j/m}$ should be determined from discrete nodes $\mathbf{x}_k^{(m)}$ ($k = 0, 1, 2, \ldots, mN$) with $mN + 1 \geq 2M + 1$. For $M = mN/2$, the node points $\mathbf{x}_k^{(m)}$ on the period-m motion can be expressed for $t_k \in [0, mT]$,

$$\mathbf{x}^{(m)}(t_k) \equiv \mathbf{x}_k^{(m)} = \mathbf{a}_0^{(m)} + \sum_{j=1}^{mN/2} \mathbf{b}_{j/m} \cos\left(\frac{j}{m}\Omega t_k\right) + \mathbf{c}_{j/m} \sin\left(\frac{j}{m}\Omega t_k\right)$$

$$= \mathbf{a}_0^{(m)} + \sum_{j=1}^{mN/2} \mathbf{b}_{j/m} \cos\left(\frac{j}{m}\frac{2k\pi}{N}\right) + \mathbf{c}_{j/m} \sin\left(\frac{j}{m}\frac{2k\pi}{N}\right)$$

$$(k = 0, 1, \ldots, mN - 1) \tag{4.35}$$

where

$$T = \frac{2\pi}{\Omega} = N\Delta t; \quad \Omega t_k = \Omega k \Delta t = \frac{2k\pi}{N}$$

$$\mathbf{a}_0^{(m)} = \frac{1}{N} \sum_{k=0}^{mN-1} \mathbf{x}_k^{(m)},$$

$$\left.\begin{array}{l} \mathbf{b}_{j/m} = \dfrac{2}{mN} \displaystyle\sum_{k=1}^{mN-1} \mathbf{x}_k^{(m)} \cos\left(k\dfrac{2j\pi}{mN}\right), \\[3ex] \mathbf{c}_{j/m} = \dfrac{2}{mN} \displaystyle\sum_{k=1}^{mN-1} \mathbf{x}_k^{(m)} \sin\left(k\dfrac{2j\pi}{mN}\right) \end{array}\right\} \quad (j = 1, 2, \ldots, mN/2) \tag{4.36}$$

and

$$\begin{aligned} \mathbf{a}_0^{(m)} &= (a_{01}^{(m)}, a_{02}^{(m)}, a_{03}^{(m)}, a_{04}^{(m)})^{\mathrm{T}}, \\ \mathbf{b}_{j/m} &= (b_{j/m1}, b_{j/m2}, b_{j/m3}, b_{j/m4})^{\mathrm{T}}, \\ \mathbf{c}_{j/m} &= (c_{j/m1}, c_{j/m2}, c_{j/m3}, c_{j/m4})^{\mathrm{T}}. \end{aligned} \tag{4.37}$$

The harmonic amplitudes and harmonic phases for the period-m motions are expressed by

$$A_{(s)j/m1} = \sqrt{b_{(s)j/m}^2 + c_{(s)j/m}^2}, \quad \varphi_{(s)j/m1} = \arctan\frac{c_{(s)j/m}}{b_{(s)j/m}},$$

$$(s = 1, 2, 3, 4). \tag{4.38}$$

Thus, the approximate expression of period-m motions in Eq. (4.33) becomes

$$\mathbf{x}^{(m)}(t) \approx \mathbf{a}_0^{(m)} + \sum_{j=1}^{mN/2} \mathbf{b}_{j/m} \cos\left(\frac{j}{m}\Omega t\right) + \mathbf{c}_{j/m} \sin\left(\frac{j}{m}\Omega t\right). \tag{4.39}$$

To reduce illustrations, only frequency-amplitude curves of the linear and angular displacements $x_1^{(m)}(t)$ and $x_2^{(m)}(t)$ for period-m motions are presented. However, the frequency-amplitudes for $y_1^{(m)}(t)$ and $y_2^{(m)}(t)$ can also be done in a similar fashion. Thus the pendulum displacement for period-m motion is given as

$$x_\alpha^{(m)}(t) \approx a_{(\alpha)0}^{(m)} + \sum_{j=1}^{mN/2} b_{(\alpha)j/m} \cos\left(\frac{j}{m}\Omega t\right) + c_{(\alpha)j/m} \sin\left(\frac{j}{m}\Omega t\right),$$

$$(\alpha = 1, 2) \tag{4.40}$$

or

$$x_\alpha^{(m)}(t) \approx a_{(\alpha)0}^{(m)} + \sum_{j=1}^{mN/2} A_{(\alpha)j/m} \cos\left(\frac{j}{m}\Omega t - \varphi_{(\alpha)j/m}\right), \tag{4.41}$$

where

$$A_{(\alpha)j/m} = \sqrt{b_{(\alpha)j/m}^2 + c_{(\alpha)j/m}^2}, \quad \varphi_{(\alpha)j/m} = \arctan\frac{c_{(\alpha)j/m}}{b_{(\alpha)j/m}}. \tag{4.42}$$

References

1. Guo Y, Luo ACJ (2022) Period-3 motions to chaos in a periodically forced nonlinear-spring pendulum. AIP Chaos, 32, 103129
2. Luo ACJ, Yuan YG (2020) Bifurcation trees of period-1 to period-2 motions in a periodically excited nonlinear spring pendulum. Journal of Vibration Testing and System Dynamics, 4(3): 201–248.
3. Guo Y (2022) Bifurcations and harmonic responses of period-1 motions in a periodically excited spring pendulum. Journal of Vibration Testing and System Dynamics, 6(3): 297–315.
4. Luo ACJ (2015) Discretization and Implicit Mapping Dynamics, HEP/Springer, Beijing/Dordrecht

Period-1 Motions to Chaos

5.1 Bifurcation Trees

As in Luo and Yuan [1], without loss of generality, a set of arbitrary parameters are chosen as

$$k_1 = 5, \quad k_2 = 100, \quad m_1 = 1, \quad c = 0.1, \quad Q_0 = 20.0,$$
$$L = 2, \quad T = 2\pi/\Omega, \quad g = 9.8. \tag{5.1}$$

The prediction of the periodic discrete nodes for stable and unstable period-1 to period-2 motions is presented in Fig. 5.1. The corresponding stability and bifurcation of the period-1 to period-2 motions are determined through eigenvalue analysis. The spring displacement x_1 and velocity y_1, angular displacement x_2 and velocity y_2 of the periodic nodes of the period-1 to period-2 motions on the Poincare mapping section are presented for $\Omega \in (0, 18)$ in Fig. 5.2a–d, respectively. The stable and unstable period-1 to period-2 motions are represented by the solid and dash curves, respectively. The acronyms 'SN', 'NB', and 'PD' are for saddle-node, Neimark and period-doubling bifurcations respectively. The multiple branches of period-1 motion to period-2 motions coexist in the periodically excited spring pendulum. For low excitation frequency, the periodic motions become more complicated. For high excitation frequency, the bifurcation points can be easily observed. The bifurcation points are listed in Tables 5.1, 5.2 and 5.3.

5.2 Frequency-Amplitude Characteristics

The discrete nodes of period-1 and period-2 motions of the nonlinear spring pendulum are obtained. From the discrete Fourier series, the nonlinear frequency-amplitude characteristics of period-1 and period-2 motions can be developed. The global views of harmonic

Y. Guo and A. C. J. Luo, *Periodic Motions to Chaos in a Spring-Pendulum System*,
Synthesis Lectures on Mechanical Engineering,
https://doi.org/10.1007/978-3-031-17883-2_5

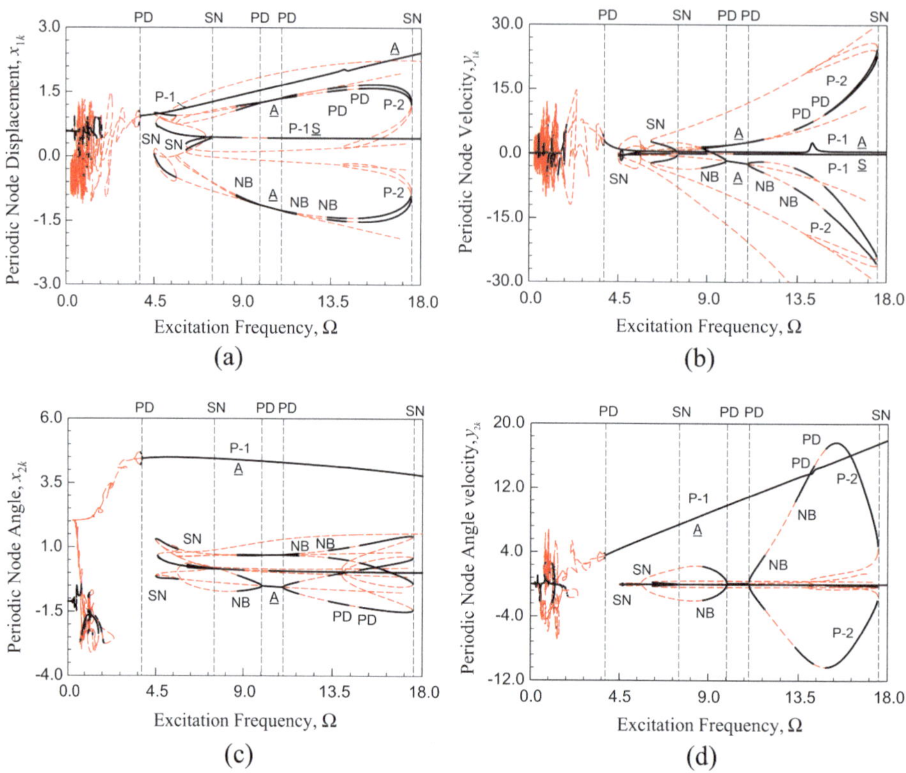

Fig. 5.1 The global view of bifurcation tree of period-1 to period-2 motions varying with excitation frequency ($\Omega \in (0, 18)$). **a** Node displacement $x_{1,k}$, **b** node velocity y_1, **c** node angle x_1, **d** node angular velocity y_2 ($k_1 = 5$, $k_2 = 100$, $m_1 = 1$, $c = 0.1$, $Q_0 = 20.0$, $L = 2$, $T = 2\pi/\Omega$). SN: saddle-node, PD: period-doubling, NB: Neimark bifurcation

amplitudes ($a_{(1)0}, A_{(1)j}$ ($j = 1, 2, 3, 15, 16$)) and ($a_{(2)0}, A_{(2)j}$ ($j = 1, 2, 3, 15, 16$)) are presented in Figs. 5.2 and 5.3, respectively.

For symmetric period-1 motions, $a_{(1)0} \neq 0$, $a_{(2)0} = 0$. For asymmetric period-1 motions, $a_{(1)0} \neq 0$, $a_{(2)0} \neq 0$. For the asymmetric period-2 motion, $a_{(1)0}^{(2)} \neq 0$ and $a_{(2)0}^{(2)} \neq 0$. The constant terms $a_{(1)0}^{(2)}$ and $a_{(2)0}^{(2)}$ varying with excitation frequency are presented in Figs. 5.2a and 5.3a, respectively. The spring oscillator in the spring-pendulum possesses dynamical behaviors different from the pendulum. This is because the nonlinear spring pendulum is a parametric system. In Fig. 5.2b–d, the first, second and third primary harmonic amplitudes are presented for the spring oscillator in the spring pendulum. However, the first, second and third primary harmonic amplitudes are presented in Fig. 5.3b–d for the pendulum oscillator in the spring pendulum system. The even terms of the harmonic-amplitude of displacement x_1 have similar patterns with the odd terms of the harmonic-amplitude of angle x_2. The first and third harmonic amplitudes for the

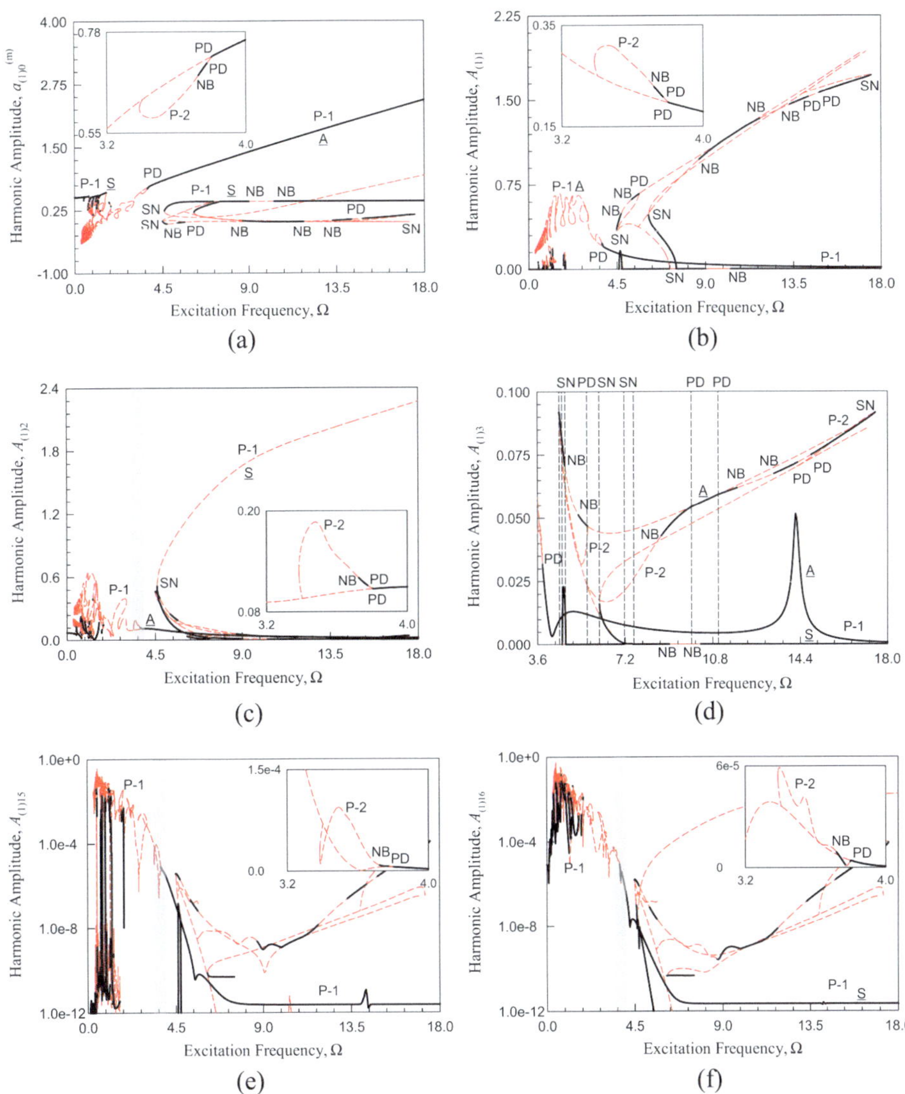

Fig. 5.2 The global view of frequency-amplitude characteristics of bifurcation tree of period-1 to period-2 motions varying with excitation frequency of $\Omega \in (0, 18)$ for displacement x_1: **a** $a_{(1)0}$, **b–f** $A_{(1)j}$ ($j = 1, 2, 3, 15, 16$) ($m = 2$, $k_1 = 5$, $k_2 = 100$, $m_1 = 1$, $c = 0.1$, $Q_0 = 20.0$, $L = 2$, $T = 2\pi/\Omega$). SN: saddle-node, PD: period-doubling, NB: Neiamrk bifurcation, P-1: period-1 motion, P-2: period-2 motion

Table 5.1 Bifurcations for symmetric period-1 motions ($k_1 = 5$, $k_2 = 100$, $c = 0.1$, $Q_0 = 20.0$, $L = 2$, $T = 2\pi/\Omega$)

	Ω	Bifurcations
1st bifurcation tree	4.744	SN (A)
	7.17	SN (A)
	7.51	SN (A)
	4.61	SN (A)
	4.589	SN (J)
	9.03	NB
	10.29	NB
2nd bifurcation tree	3.8	PD
3rd bifurcation tree	1.755	SN(A)
	1.58	NB
	1.63	NB
	1.4565	SN(J)
	1.851	SN(A)
	1.91	NB
4th bifurcation tree	1.139	SN(A)
	1.233	SN(A)
	1.23	SN(A)
	1.266	SN(J)
	1.133	SN(J)
	1.1463	SN(A)
	1.229	SN(A)
	1.621	SN(J)
	1.3159	SN(J)
	1.3187	NB
	0.937	SN(A)
	0.9244	SN(A)
	0.902	SN(A)
	0.9556	SN(A)
	0.9522	NB
	0.9435	SN(J)
	0.762	SN(A)

(continued)

Table 5.1 (continued)

	Ω	Bifurcations
	0.704	SN(A)
	0.7023	SN(J)
	0.6993	SN(J)
	0.6954	NB
	0.6942	SN(A)
	0.6872	SN(A)
	0.6857	SN(J)
	0.695	SN(J)
	0.7471	SN(A)
	0.7428	NB
	0.6667	SN(A)
	0.6657	NB
	0.7313	SN(J)
	0.488	SN(A)
	0.4748	SN(A)
	0.467	SN(A)
	0.4725	SN(J)
	0.4644	SN(J)
	0.4718	SN(J)
	0.4605	SN(J)

Note J: jumping phenomena, A: from symmetric to asymmetric period-1 motions, NB: Neimark Bifurcation, SN: saddle-node bifurcation, PD: period-doubling

spring oscillator still have the dynamical behaviors of the parametric hardening Duffing oscillator. The zoomed views of frequency-amplitude characteristics are presented in Figs. 5.4, 5.5, 5.6, 5.7, 5.8, 5.9, 5.10, 5.11, 5.12, 5.13, 5.14, 5.15, 5.16, 5.17, 5.18 and 5.19, which are tabulated in Table 5.4.

The zoomed view of frequency-amplitude characteristics for $\Omega \in (0, 2.25)$ are presented in Figs. 5.4 and 5.5. In Fig. 5.4a–f, the frequency-amplitude characteristics of the nonlinear spring system are presented through $a_{(1)0}^{(m)}$ and $A_{(1)(2j)/2} = A_{(1)j}$ ($j = 1, 2, 3, 15, 16$). For the parametric motions of the spring oscillator, for the symmetric period-1 motions, $A_{(1)(2l-1)} = 0$ ($l = 1, 2, \ldots$) but $A_{(1)2l} \neq 0$. In Fig. 5.5a–f, the frequency-amplitude characteristics of the pendulum system are presented through $a_{(2)0}^{(m)}$ and $A_{(2)2j/2} = A_{(2)j}$ ($j = 1, 2, 3, 15, 16$). For the symmetric period-1 motions, $a_{(2)0} = 0$ and $A_{(2)2l} = 0$ ($l = 1, 2, \ldots$) but $A_{(2)(2l-1)} \neq 0$. If $\lim\limits_{j \to \infty} A_{(2)j/m} \neq 0$, the travelable

Table 5.2 Bifurcations for asymmetric period-1 motion ($k_1 = 5$, $k_2 = 100$, $c = 0.1$, $Q_0 = 20.0$, $L = 2$, $T = 2\pi/\Omega$)

	Ω	Bifurcations
1st bifurcation tree	4.49	SN(J)
	4.486	PD
	6.14	SN(J)
	5.66	PD
	9.93	PD
	11.03	PD
2nd bifurcation tree	3.807	PD
3rd bifurcation tree	1.774	PD
	1.8495	PD
4th bifurcation tree	1.1032	NB
	1.1036	PD
	1.0861	NB
	1.0866	NB
	1.0882	NB
	1.0901	PD
	1.1502	PD
	1.1649	SN(J)
	1.1354	SN(J)
	1.1389	PD
	1.1502	PD
	1.1036	PD
	1.2044	PD
	1.099	PD
	0.4862	NB
	0.4837	NB
	0.4826	PD
	0.4681	PD
	0.4679	SN(A)
	0.46777	SN(J)
	0.931	NB
	0.7546	PD

(continued)

Table 5.2 (continued)

	Ω	Bifurcations
	0.706	PD
	0.7472	PD
	0.6914	PD
	0.6882	PD
	0.9017	PD
	0.6668	PD

Note J: jumping phenomena, A: from symmetric to asymmetric period-1 motions. PD: period-doubling from period-1 to period-2 motion, SN: saddle-node bifurcation for onset of asymmetric period-1 motion, NB: Neimark bifurcation

periodic motions of the pendulum system exist. For the travelable periodic motions in the pendulum, the harmonic amplitudes will approach zero if the harmonic order approaches infinity. For non-travelable periodic motions in the pendulum, $\lim_{j \to \infty} A_{(2)j/m} = 0$. Thus, harmonic amplitudes of the pendulum for high frequency have $A_{(2)15} \sim 10^{-12}$ and $A_{(2)16} \sim 10^{-12}$ for the non-travelable periodic motion. For the periodic motions of the nonlinear spring system, $\lim_{j \to \infty} A_{(1)j/m} = 0$ whether the correlated pendulum has the non-travelable or travelable periodic motions. The harmonic amplitudes $A_{(1)15}$ and $A_{(1)16}$ for period-1 and period-2 motions have about the quantity levels of $\varepsilon = 10^{-12}$. The harmonic amplitudes $A_{(2)15}$ and $A_{(2)16}$ for the non-travelable period-1 and period-2 motions of the pendulum have about the quantity levels of $\varepsilon = 10^{-12}$. However, the harmonic amplitudes $A_{(2)15}$ and $A_{(2)16}$ for the travelable period-1 and period-2 motions of the pendulum have about the quantity levels of $\varepsilon = 10^{-1}$ because the Fourier series cannot work for the travelable periodic motions in the pendulum systems.

The further zoomed views of frequency-amplitude characteristics for $\Omega \in (0.45, 0.495)$ are presented in Figs. 5.6 and 5.7 for the non-travelable periodic motions of the spring-pendulum systems. In Fig. 5.6a–f, the frequency-amplitude characteristics of the nonlinear spring system are also presented through $a_{(1)0}^{(m)}$ and $A_{(1)(2j)/m} = A_{(1)j}$ ($j = 1, 2, 3, 15, 16$; $m = 2$) for period-1 and period-2 motions. For the symmetric period-1 motions, $A_{(1)(2l-1)} = 0$ ($l = 1, 2, \ldots$) are clearly observed. For the asymmetric period-1 and period-2 motions, $a_{(1)0} \neq 0$ and $A_{(1)(2l-1)} \neq 0$ with $A_{(1)(2l)} \neq 0$ are obtained. The frequency-amplitude characteristics of the pendulum system are placed in Fig. 5.8a–f through $a_{(2)0}^{(m)}$ and $A_{(2)(2j)/m} = A_{(2)j}$ ($j = 1, 2, 3, 15, 16$; $m = 2$). For the symmetric period-1 motions, $a_{(1)0}^{(1)} = 0$ and $A_{(2)(2l)} = 0$ ($l = 1, 2, \ldots$) are also obtained. For the asymmetric period-1 and period-2 motions, $a_{(2)0}^{(m)} \neq 0$ and $A_{(2)(2l)/m} \neq 0$ with $A_{(1)(2l-1)} \neq 0$ are obtained. For the symmetric period-1 motions, the solutions of the parametric-spring oscillator is

Table 5.3 Bifurcations for period-2 motions ($\alpha_1 = 10.0$, $\alpha_2 = 5.0$, $\beta = 10.0$, $\delta = 0.5$, $Q_0 = 200$, $T = 2\pi/\Omega$)

	Ω	Bifurcations
1st bifurcation tree	11.031	SN
	8.69	NB
	11.83	NB
	14.85	PD
	14.3	PD
	13.311	NB
	13.9	SN
2nd bifurcation tree	3.807	SN
	3.8037	PD
	3.7765	PD
	3.7245	NB
	3.4025	SN
3rd bifurcation tree	1.774	SN
	1.777	NB
	1.849	SN
4th bifurcation tree	0.482	NB
	0.4684	PD
	1.1036	PD
	1.1167	NB
	1.0898	PD
	1.0828	PD
	1.1758	PD
	1.1782	SN
	0.6668	PD
	0.6893	PD
	0.6907	PD
	0.7535	PD
	0.7052	PD
	0.7472	NB
	0.9016	NB

Note SN for onset of asymmetric period-2 motions, *PD* period-doubling from period-2 to period-4 motion. *SN* saddle-node bifurcation between unstable and stable period-1 motions

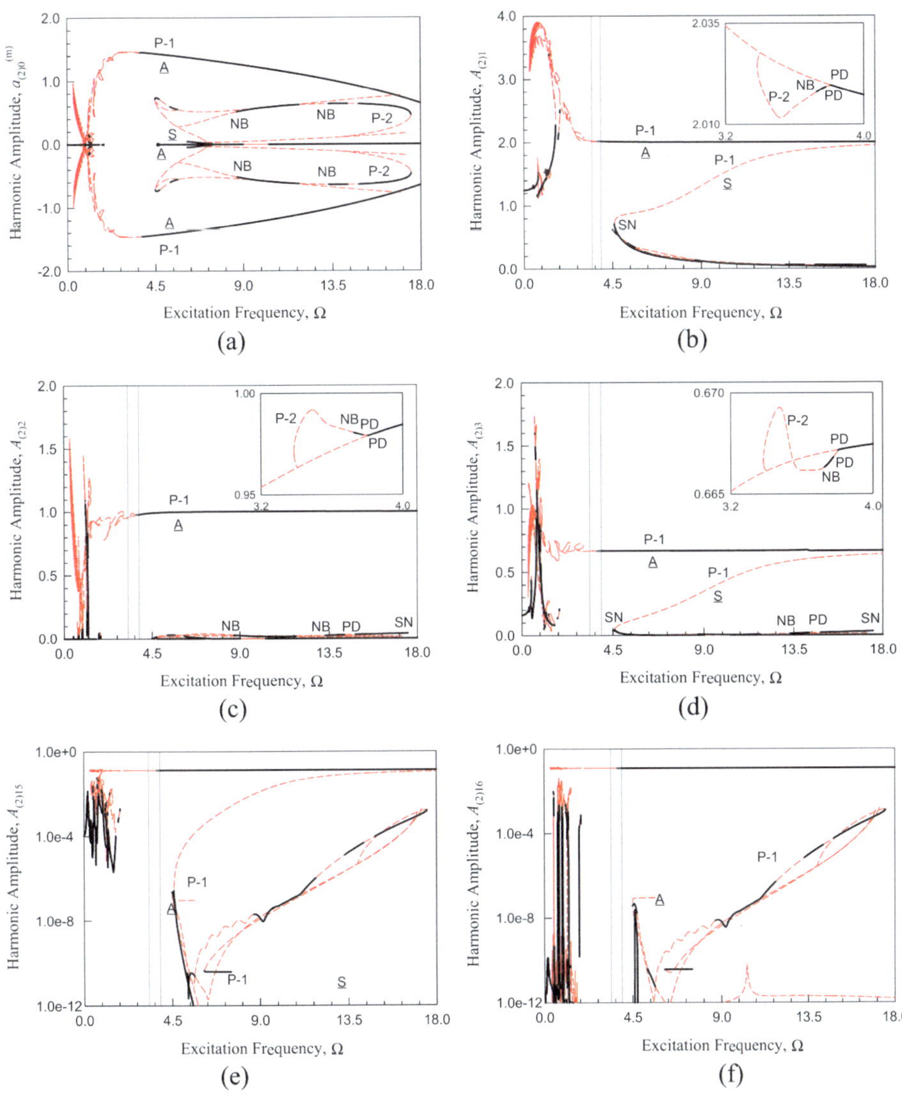

Fig. 5.3 The global view of frequency-amplitude characteristics of bifurcation tree of period-1 to period-2 motions varying with excitation frequency of $\Omega \in (0, 18)$ for displacement x_2: **a** $a_{(2)0}$, **b**–**f** $A_{(2)j}$ ($k = 1, 2, 3, 15, 16$) ($m = 2$, $k_1 = 5$, $k_2 = 100$, $m_1 = 1$, $c = 0.1$, $Q_0 = 20.0$, $L = 2$, $T = 2\pi/\Omega$). A: Asymmetric, S: symmetric, SN: saddle-node, PD: period-doubling, NB: Neiamrk bifurcation, P-1: period-1 motion, P-2: period-2 motion

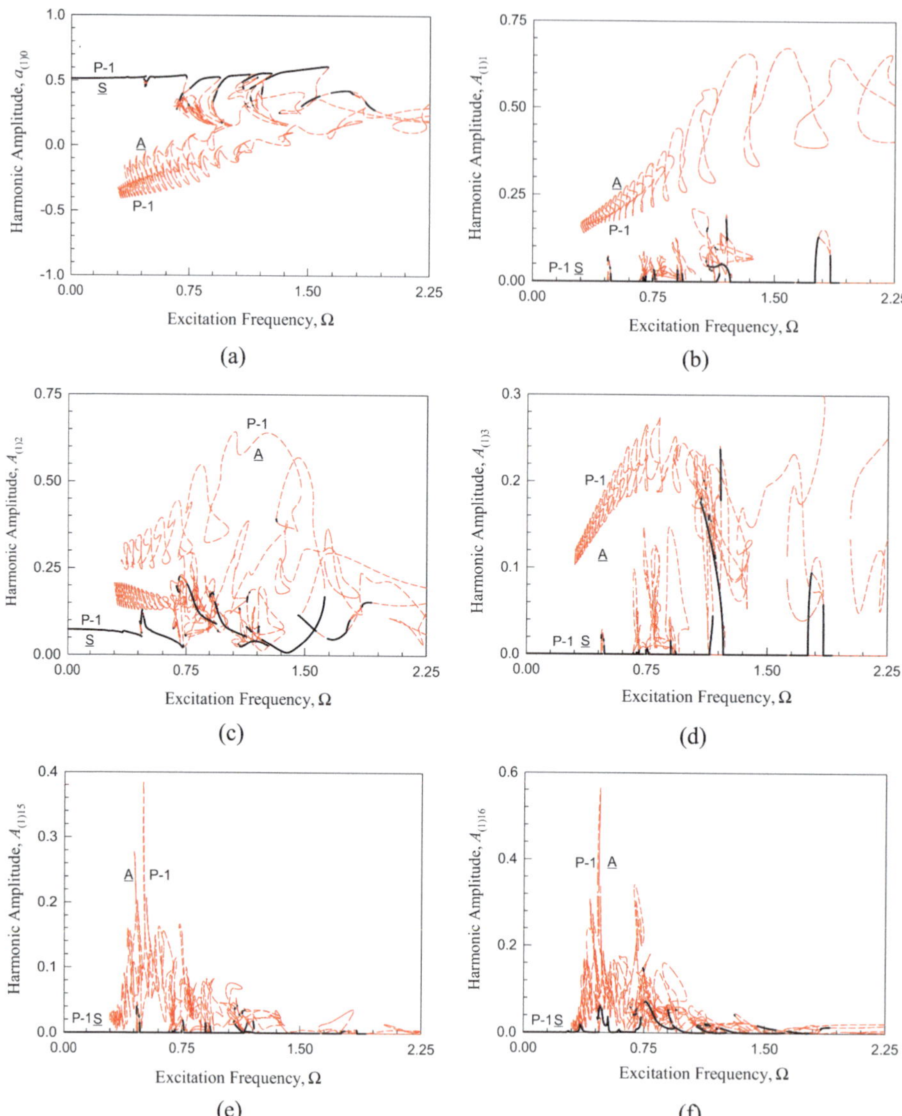

Fig. 5.4 A zoomed view of the frequency-amplitude characteristics of bifurcation tree of period-1 to period-2 motions varying with excitation frequency of $\Omega \in (0, 2.25)$ for displacement x_1: **a** $a_{(1)0}$, **b–f** $A_{(1)j}$ ($j = 1, 2, 3, 15, 16$) ($m = 2$, $k_1 = 5$, $k_2 = 100$, $m_1 = 1$, $c = 0.1$, $Q_0 = 20.0$, $L = 2$, $T = 2\pi/\Omega$). A: Asymmetric, S: symmetric, P-1: period-1 motion

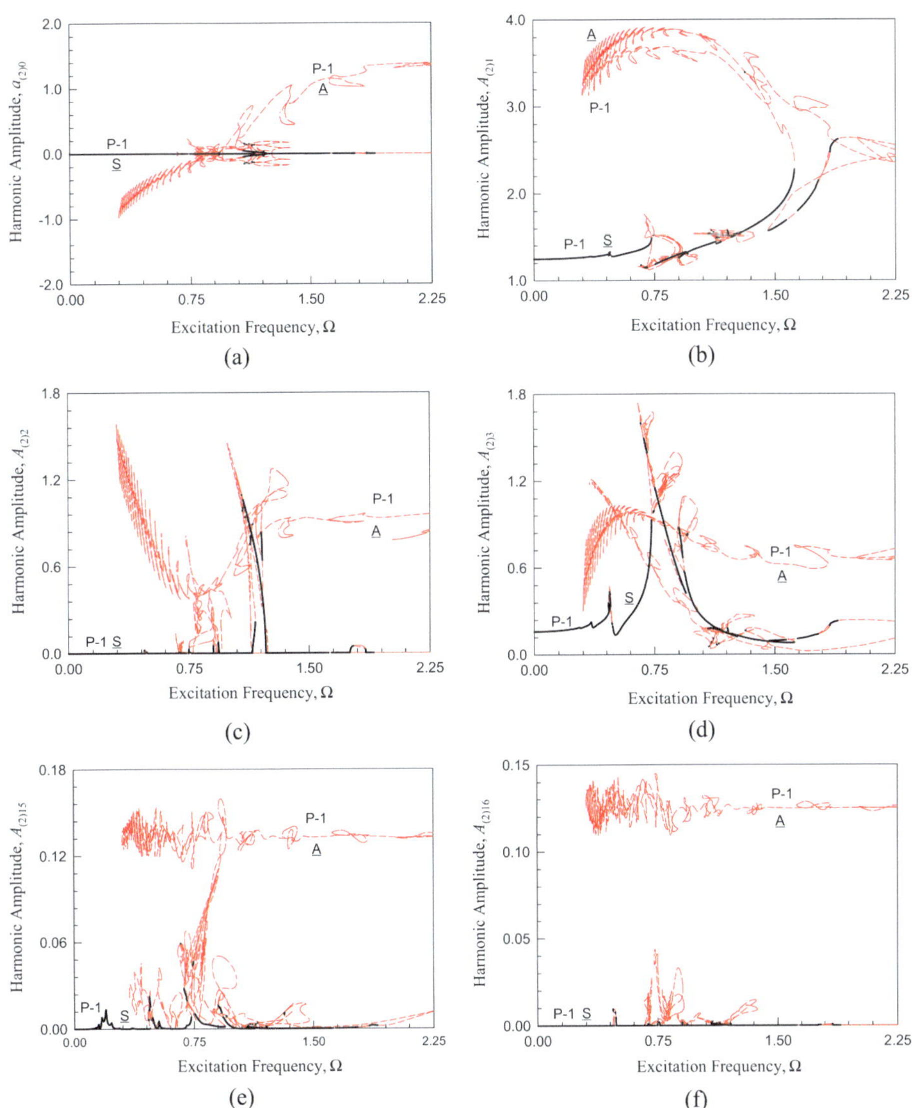

Fig. 5.5 A zoomed view of the frequency-amplitude characteristics of bifurcation tree of period-1 to period-2 motions varying with excitation frequency of $\Omega \in (0, 2.25)$ for displacement x_2: **a** $a_{(2)0}$, **b–f** $A_{(2)j}$ ($j = 1, 2, 3, 15, 16$) ($m = 2$, $k_1 = 5$, $k_2 = 100$, $m_1 = 1$, $c = 0.1$, $Q_0 = 20.0$, $L = 2$, $T = 2\pi / \Omega$). A: Asymmetric, S: symmetric, P-1: period-1 motion, P-2: period-2 motion

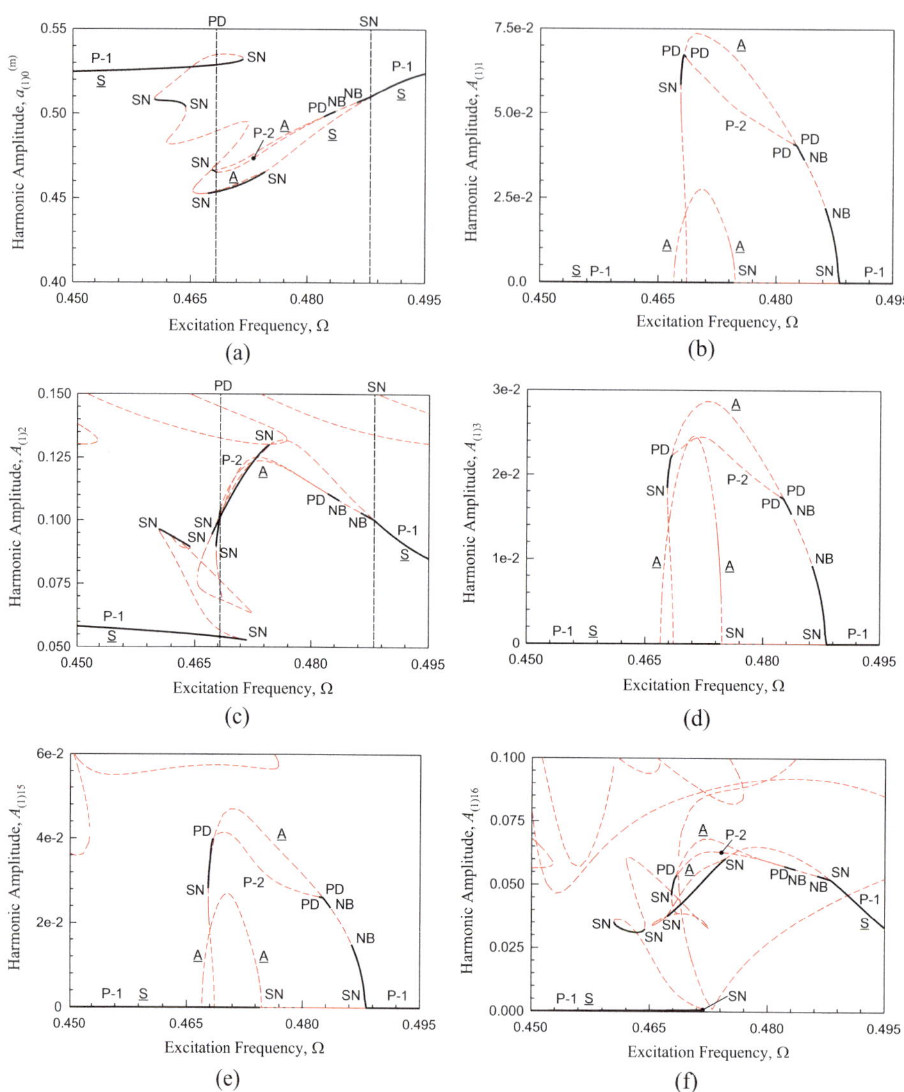

Fig. 5.6 A zoomed view of the frequency-amplitude characteristics of bifurcation tree of period-1 to period-2 motions varying with excitation frequency of $\Omega \in (0.450, 0.495)$ for displacement x_1: **a** $a_{(1)0}$, **b–f** $A_{(1)j}$ ($j = 1, 2, 3, 15, 16$) ($m = 2$, $k_1 = 5$, $k_2 = 100$, $m_1 = 1$, $c = 0.1$, $Q_0 = 20.0$, $L = 2$, $T = 2\pi/\Omega$). A: Asymmetric, S: symmetric, SN: saddle-node, PD: period-doubling, NB: Neimark bifurcation, P-1: period-1 motion, P-2: period-2 motion

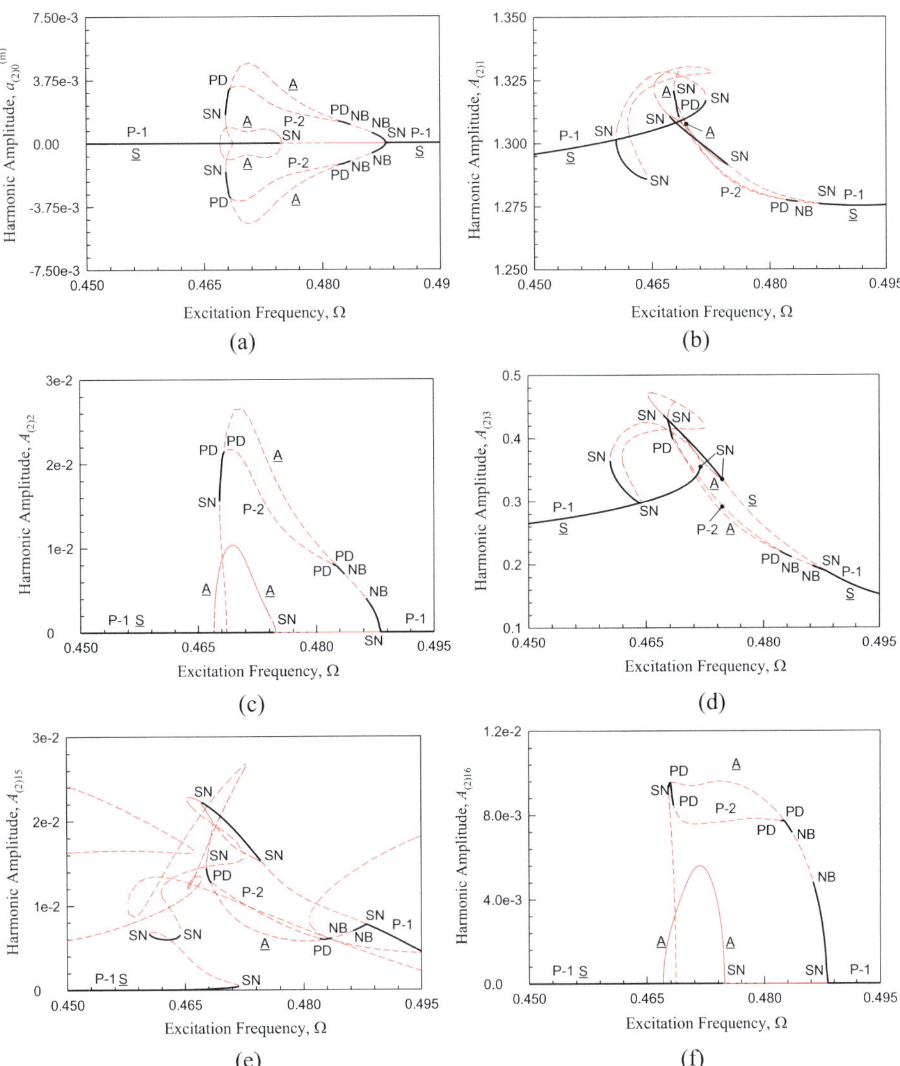

Fig. 5.7 A zoomed view of the frequency-amplitude characteristics of bifurcation tree of period-1 to period-2 motions varying with excitation frequency of $\Omega \in (0, 450.495)$ for displacement x_2: **a** $a_{(2)0}$, **b–f** $A_{(2)j}$ ($j = 1, 2, 3, 15, 16$) ($m = 2$, $k_1 = 5$, $k_2 = 100$, $m_1 = 1$, $c = 0.1$, $Q_0 = 20.0$, $L = 2$, $T = 2\pi/\Omega$). A: Asymmetric, S: symmetric, SN: saddle-node, PD: period-doubling, NB: Neimark bifurcation, P-1: period-1 motion, P-2: period-2 motion

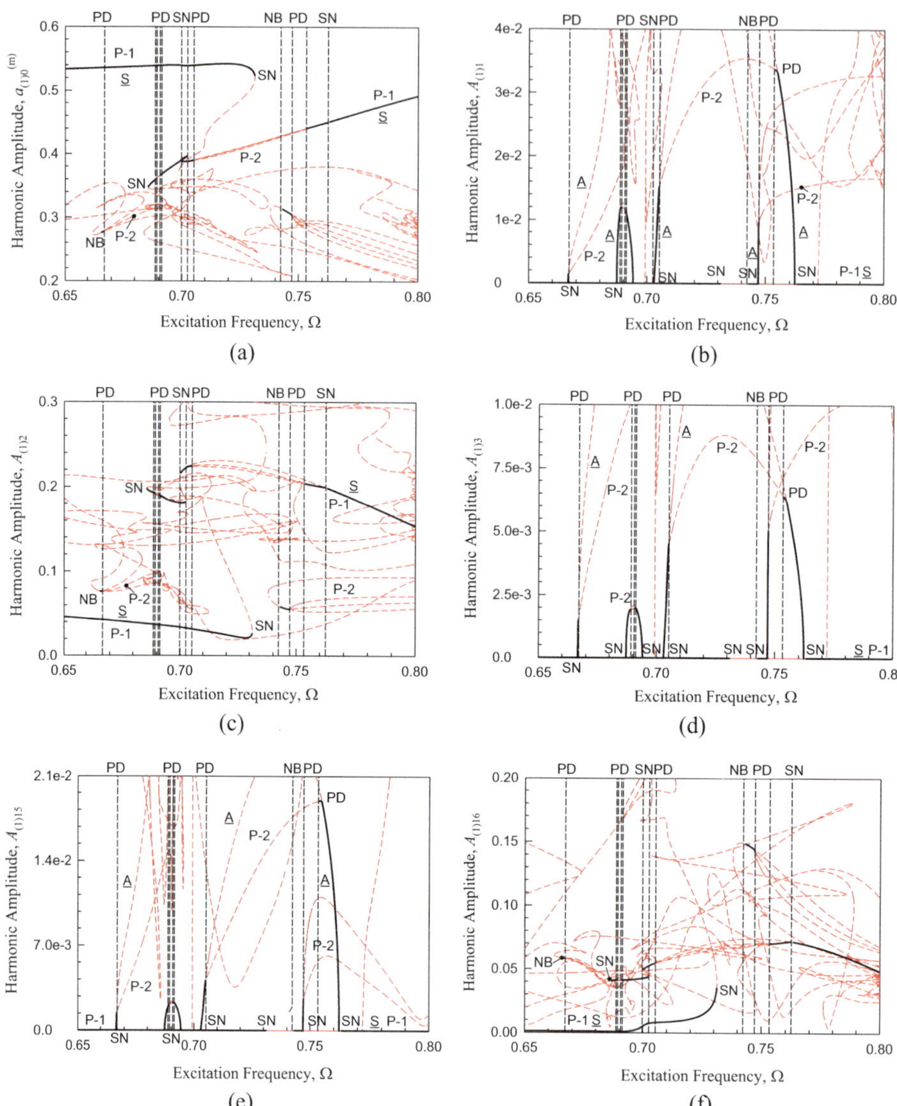

Fig. 5.8 A zoomed view of the frequency-amplitude characteristics of bifurcation tree of period-1 to period-2 motions varying with excitation frequency of $\Omega \in (0.65, 0.80)$ for displacement x_1: **a** $a_{(1)0}$, **b–f** $A_{(1)j}$ $(j = 1, 2, 3, 15, 16)$ $(m = 2, k_1 = 5, k_2 = 100, m_1 = 1, c = 0.1, Q_0 = 20.0, L = 2, T = 2\pi/\Omega)$. A: Asymmetric, S: symmetric, SN: saddle-node, PD: period-doubling, NB: Neimark bifurcation, P-1: period-1 motion, P-2: period-2 motion

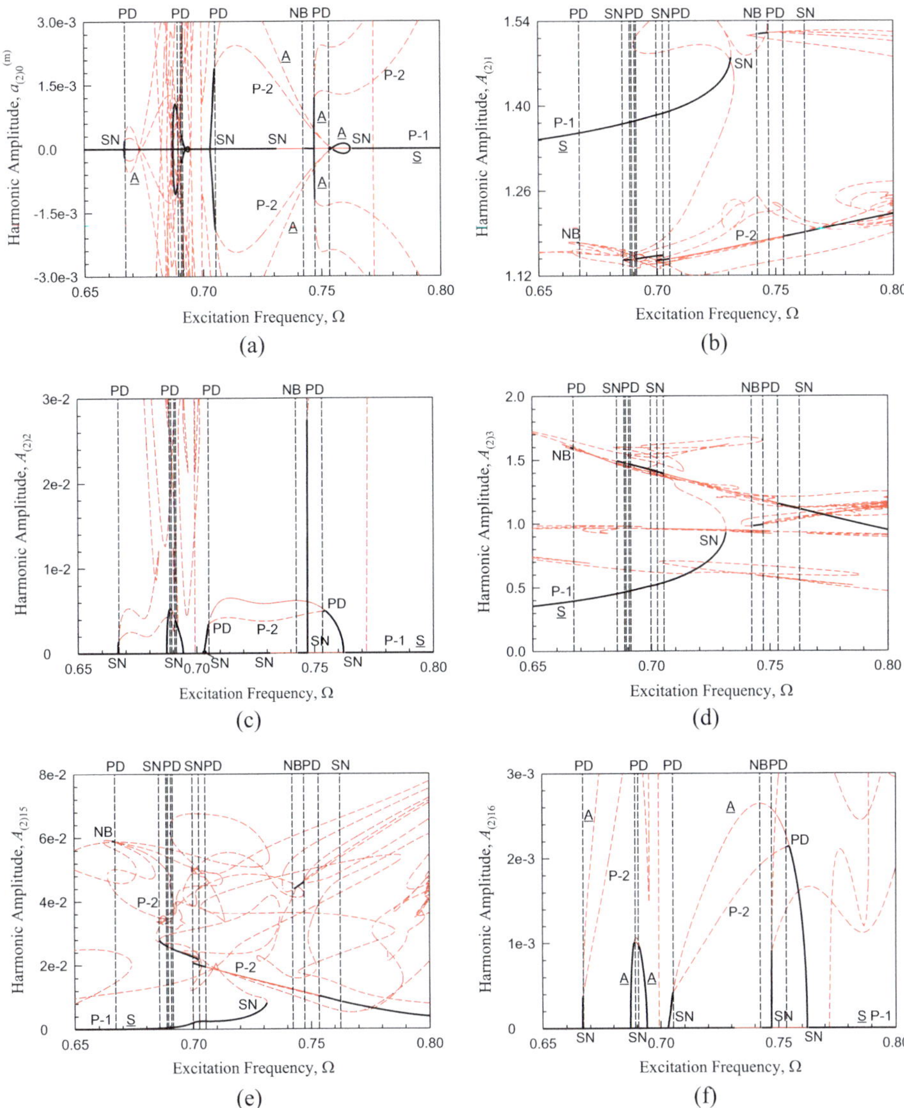

Fig. 5.9 A zoomed view of the frequency-amplitude characteristics of bifurcation tree of period-1 to period-2 motions varying with excitation frequency of $\Omega \in (0.65, 0.80)$ for displacement x_2: **a** $a_{(2)0}$, **b–f** $A_{(2)j}$ ($j = 1, 2, 3, 15, 16$) ($m = 2$, $k_1 = 5$, $k_2 = 100$, $m_1 = 1$, $c = 0.1$, $Q_0 = 20.0$, $L = 2$, $T = 2\pi/\Omega$). A: Asymmetric, S: symmetric, SN: saddle-node, PD: period-doubling, NB: Neimark bifurcation, P-1: period-1 motion, P-2: period-2 motion

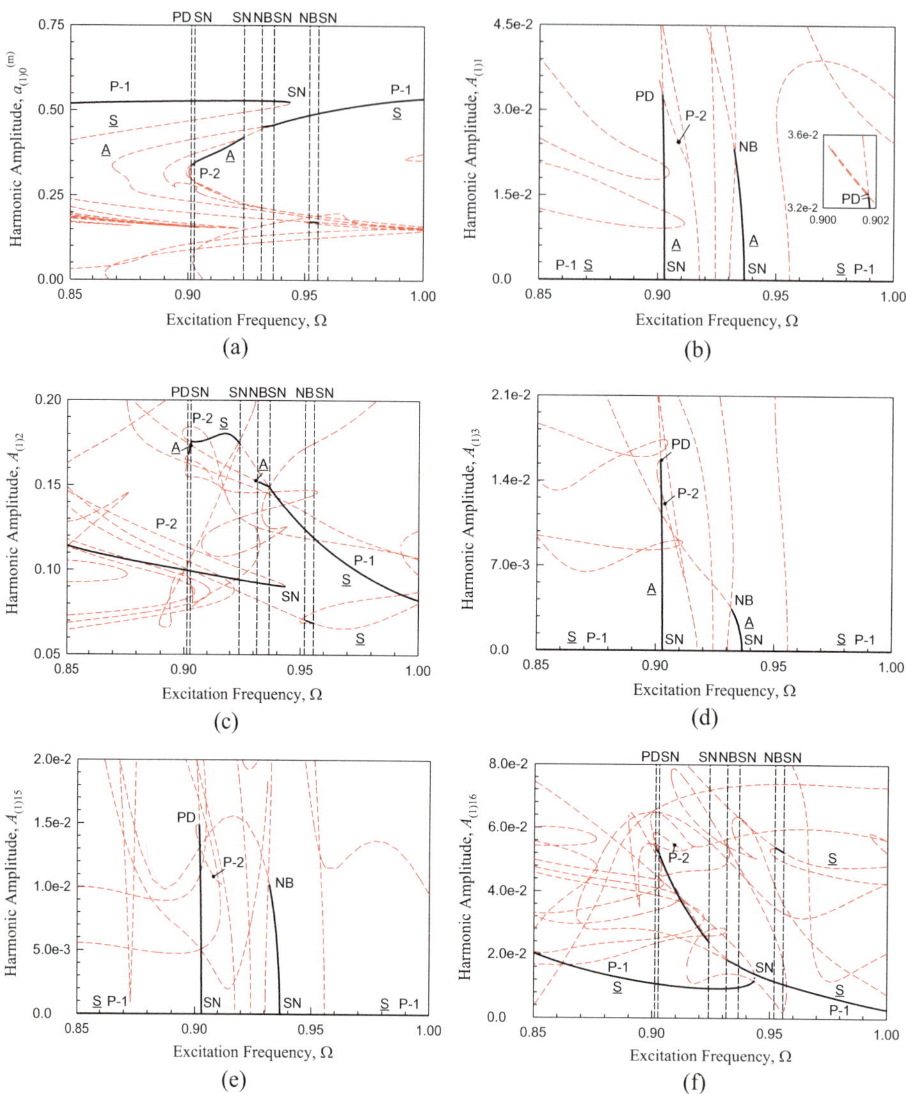

Fig. 5.10 A zoomed view of the frequency-amplitude characteristics of bifurcation tree of period-1 to period-2 motions varying with excitation frequency of $\Omega \in (0.85, 1.00)$ for displacement x_1: **a** $a_{(1)0}$, **b–f** $A_{(1)j}$ ($j = 1, 2, 3, 15, 16$) ($m = 2$, $k_1 = 5$, $k_2 = 100$, $m_1 = 1$, $c = 0.1$, $Q_0 = 20.0$, $L = 2$, $T = 2\pi/\Omega$). A: Asymmetric, S: symmetric, SN: saddle-node, PD: period-doubling, NB: Neimark bifurcation, P-1: period-1 motion, P-2: period-2 motion

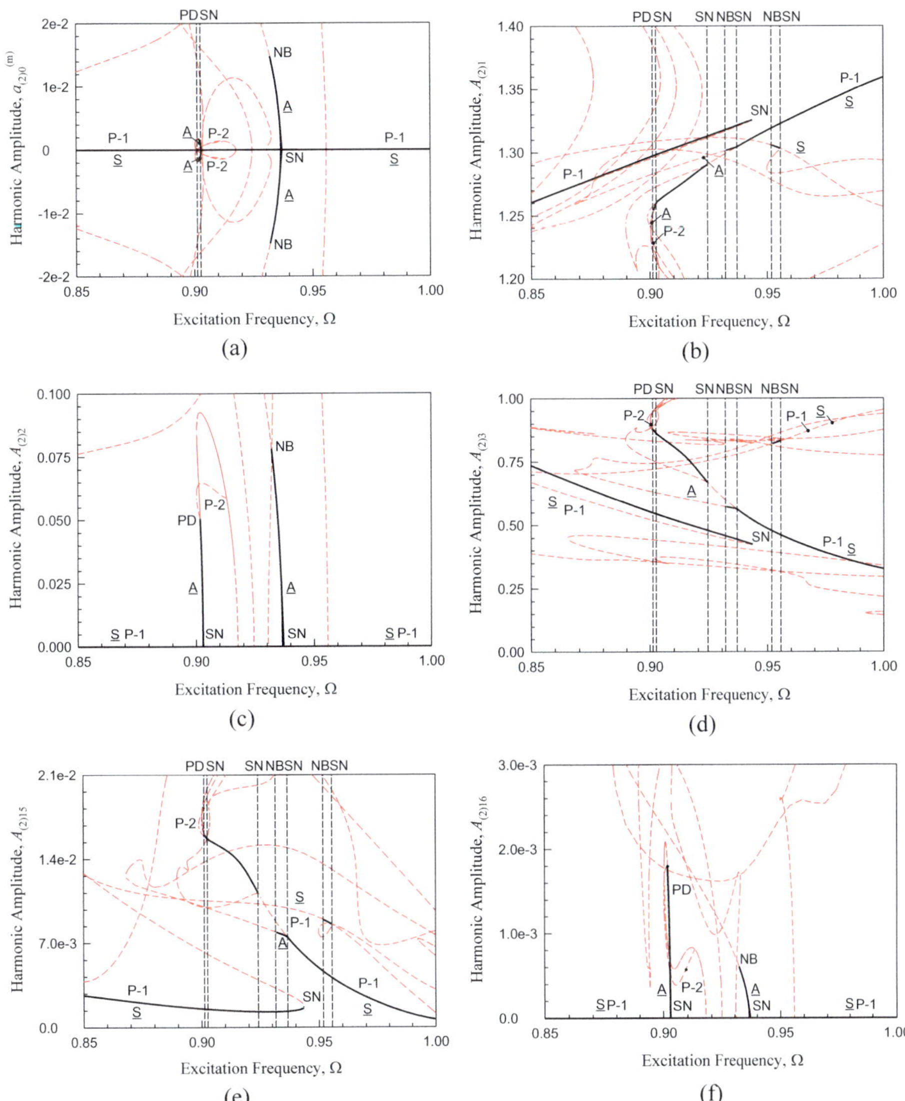

Fig. 5.11 A zoomed view of the frequency-amplitude characteristics of bifurcation tree of period-1 to period-2 motions varying with excitation frequency of $\Omega \in (0.85, 1.00)$ for displacement x_2: **a** $a_{(2)0}$, **b–f** $A_{(2)j}$ ($j = 1, 2, 3, 15, 16$) ($m = 2, k_1 = 5, k_2 = 100, m_1 = 1, c = 0.1, Q_0 = 20.0, L = 2, T = 2\pi/\Omega$). A: Asymmetric, S: symmetric, SN: saddle-node, PD: period-doubling, NB: Neimark bifurcation, P-1: period-1 motion, P-2: period-2 motion

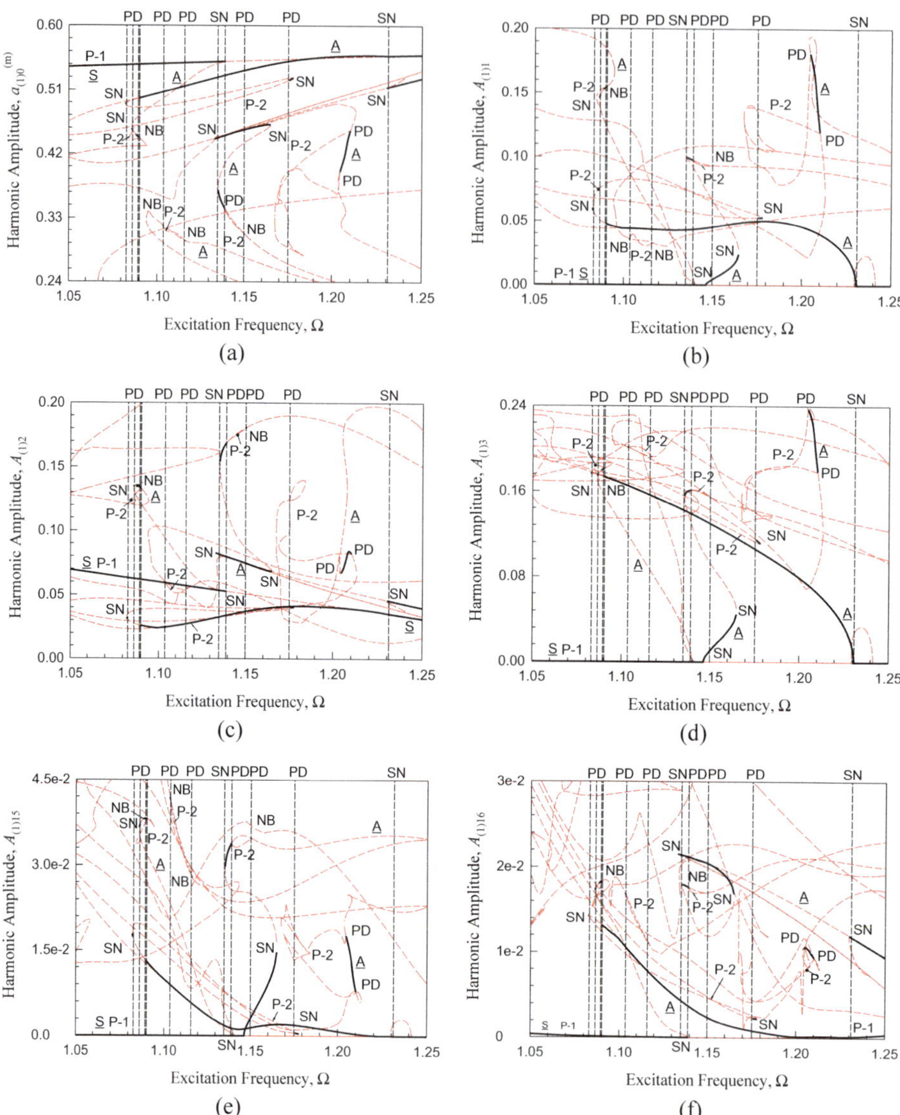

Fig. 5.12 A further zoomed view of the frequency-amplitude characteristics of bifurcation tree of period-1 to period-2 motions varying with excitation frequency of $\Omega \in (1.05, 1.25)$ for displacement x_1: **a** $a_{(1)0}$, **b–f** $A_{(1)j}$ ($j = 1, 2, 3, 15, 16$) ($m = 2$, $k_1 = 5$, $k_2 = 100$, $m_1 = 1$, $c = 0.1$, $Q_0 = 20.0$, $L = 2$, $T = 2\pi/\Omega$). A: Asymmetric, S: symmetric, SN: saddle-node, PD: period-doubling, NB: Neimark bifurcation, P-1: period-1 motion, P-2: period-2 motion

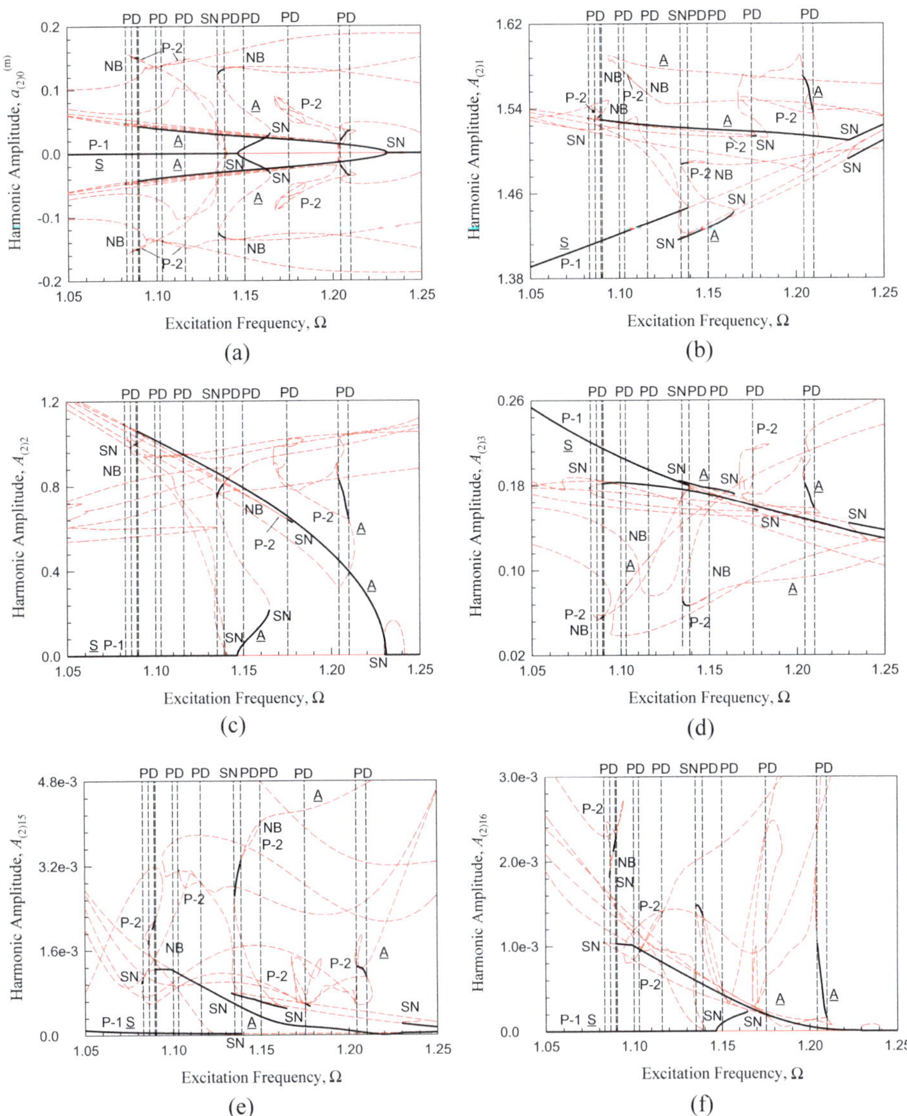

Fig. 5.13 A further zoomed view of the frequency-amplitude characteristics of bifurcation tree of period-1 to period-2 motions varying with excitation frequency of $\Omega \in (1.05, 1.25)$ for displacement x_2: **a** $a_{(2)0}$, **b–f** $A_{(2)j}$ ($j = 1, 2, 3, 15, 16$) ($m = 2$, $k_1 = 5$, $k_2 = 100$, $m_1 = 1$, $c = 0.1$, $Q_0 = 20.0$, $L = 2$, $T = 2\pi/\Omega$). A: Asymmetric, S: symmetric, SN: saddle-node, PD: period-doubling, NB: Neimark bifurcation, P-1: period-1 motion, P-2: period-2 motion

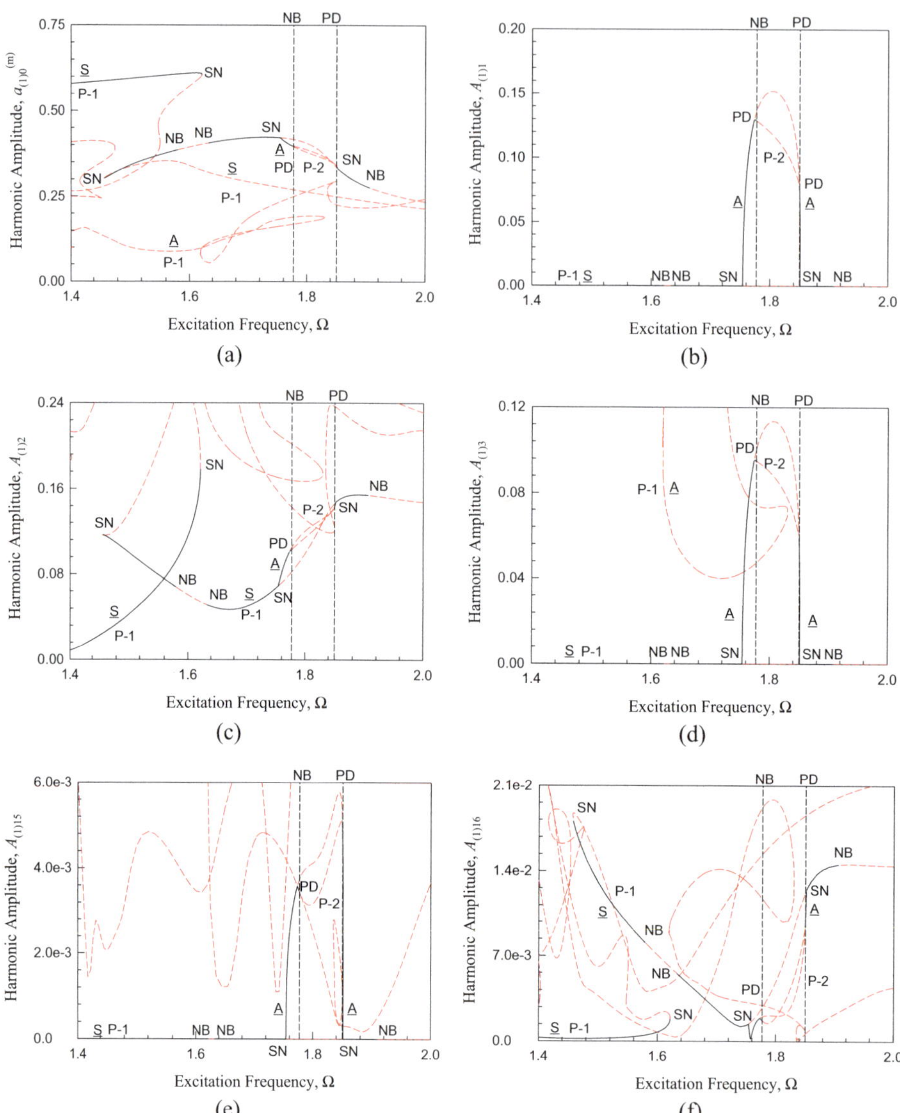

Fig. 5.14 A further zoomed view of the frequency-amplitude characteristics of bifurcation tree of period-1 to period-2 motions varying with excitation frequency of $\Omega \in (1.4, 2.0)$ for displacement x_1: **a** $a_{(1)0}$, **b–f** $A_{(1)j}$ $(j = 1, 2, 3, 15, 16)$ $(m = 2, k_1 = 5, k_2 = 100, m_1 = 1, c = 0.1, Q_0 = 20.0, L = 2, T = 2\pi/\Omega)$. A: Asymmetric, S: symmetric, SN: saddle-node, PD: period-doubling, NB: Neimark bifurcation, P-1: period-1 motion, P-2: period-2 motion

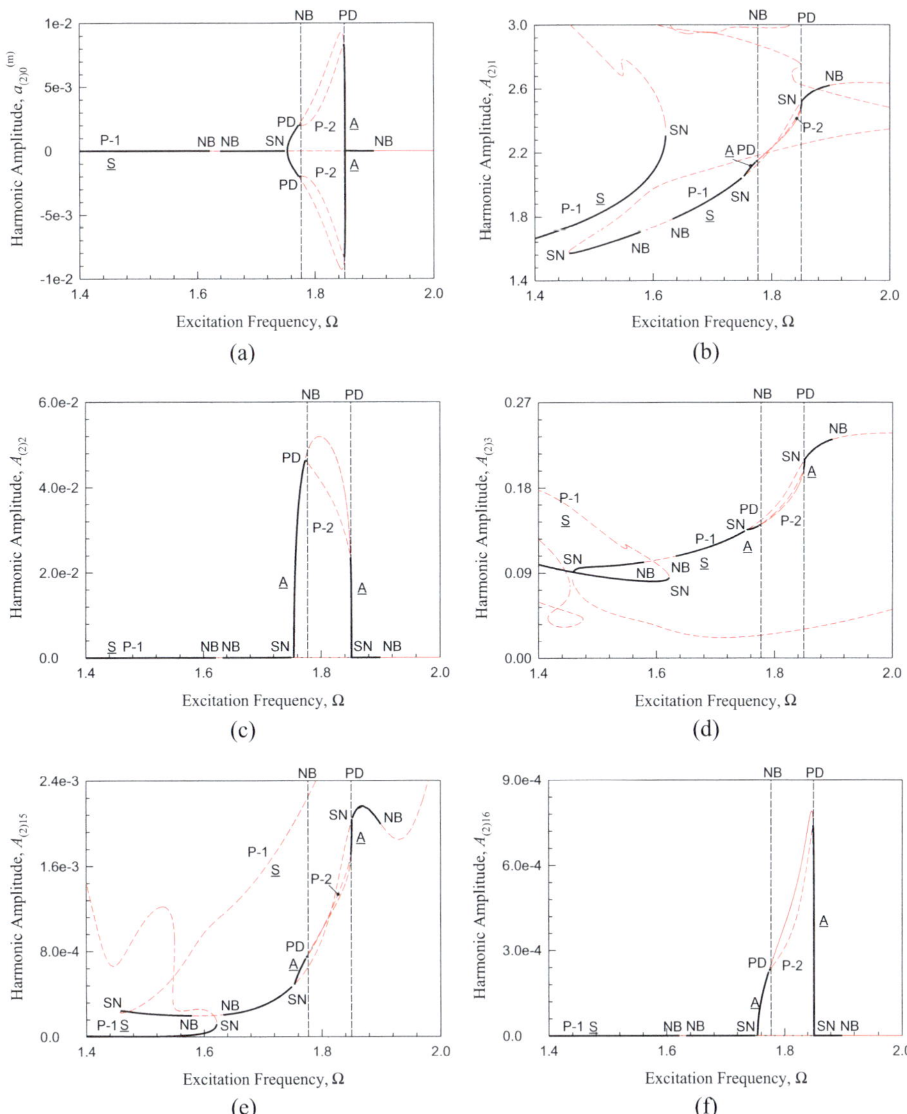

Fig. 5.15 A further zoomed view of the frequency-amplitude characteristics of bifurcation tree of period-1 to period-2 motions varying with excitation frequency ($\Omega \in (1.4, 2.0)$) for displacement x_2, **a** $a_{(2)0}$, **b**–**d** $A_{(2)j}$ ($j = 1, 2, 3, 15, 16$) ($m = 2$, $k_1 = 5$, $k_2 = 100$, $m_1 = 1$, $c = 0.1$, $Q_0 = 20.0$, $L = 2$, $T = 2\pi/\Omega$). A: Asymmetric, S: symmetric, SN: saddle-node, PD: period-doubling, NB: Neimark bifurcation, P-1: period-1 motion, P-2: period-2 motion

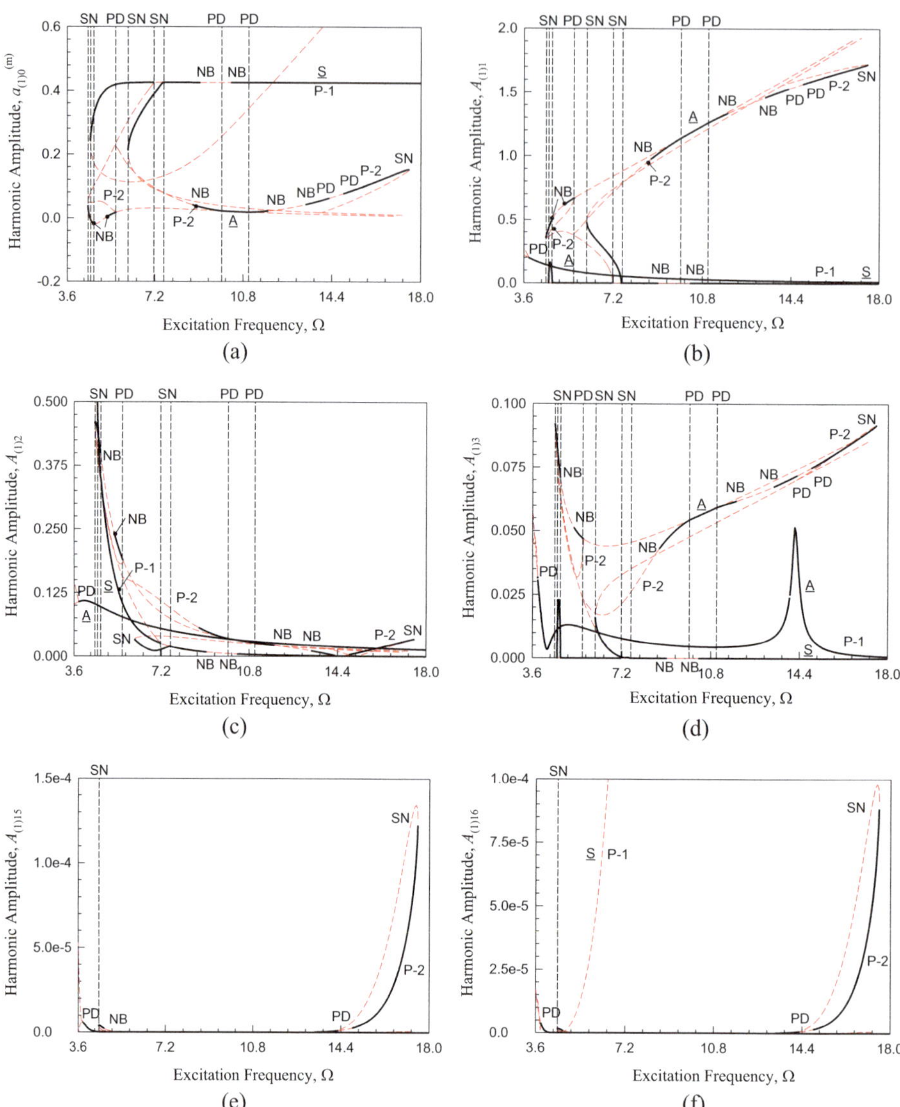

Fig. 5.16 A zoomed view of the frequency-amplitude characteristics of bifurcation tree of period-1 to period-2 motions varying with excitation frequency of $\Omega \in (3.6, 18.0)$ for displacement x_1: **a** $a_{(1)0}$, **b–f** $A_{(1)j}$ ($j = 1, 2, 3, 15, 16$) ($m = 2$, $k_1 = 5$, $k_2 = 100$, $m_1 = 1$, $c = 0.1$, $Q_0 = 20.0$, $L = 2$, $T = 2\pi/\Omega$). A: Asymmetric, S: symmetric, SN: saddle-node, PD: period-doubling, NB: Neimark bifurcation, P-1: period-1 motion, P-2: period-2 motion

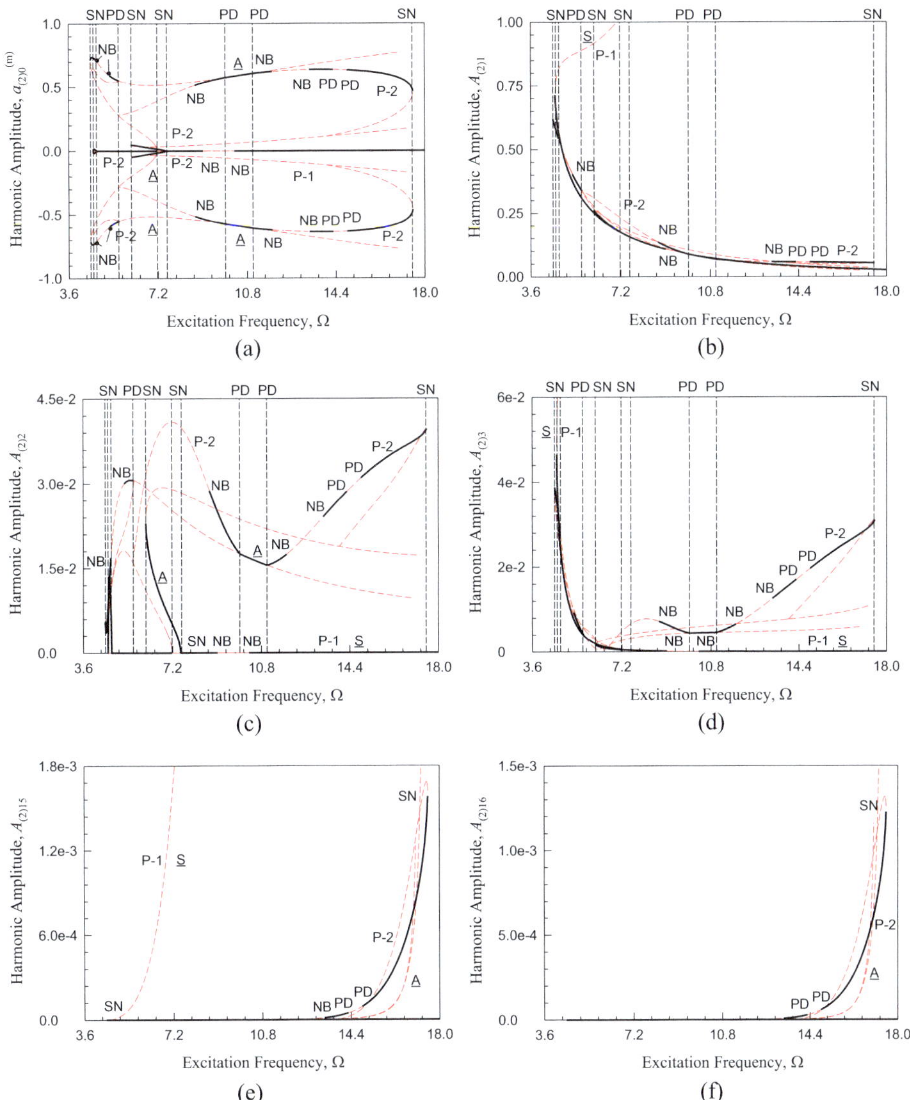

Fig. 5.17 A zoomed view of the frequency-amplitude characteristics of bifurcation tree of period-1 to period-2 motions varying with excitation frequency of $\Omega \in (3.6, 18.0)$ for displacement x_2: **a** $a_{(2)0}$, **b–d** $A_{(2)j}$ ($j = 1, 2, 3, 15, 16$) ($m = 2$, $k_1 = 5$, $k_2 = 100$, $m_1 = 1$, $c = 0.1$, $Q_0 = 20.0$, $L = 2$, $T = 2\pi/\Omega$). A: Asymmetric, S: symmetric, SN: saddle-node, PD: period-doubling, NB: Neimark bifurcation, P-1: period-1 motion, P-2: period-2 motion

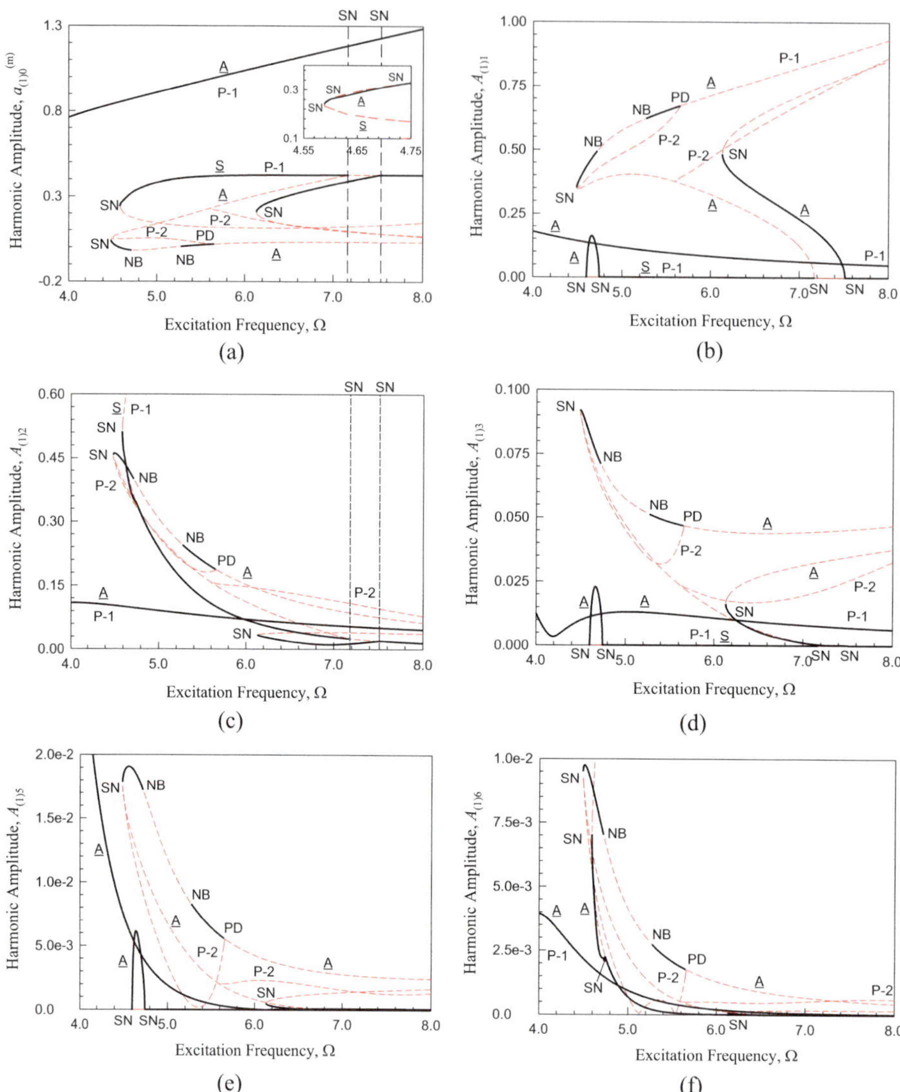

Fig. 5.18 A further zoomed view of the frequency-amplitude characteristics of bifurcation tree of period-1 to period-2 motions varying with excitation frequency of $\Omega \in (2.0, 8.0)$ for displacement x_1: **a** $a_{(1)0}$, **b–f** $A_{(1)j}$ ($j = 1, 2, 3, 5, 6$) ($m = 2$, $k_1 = 5$, $k_2 = 100$, $m_1 = 1$, $c = 0.1$, $Q_0 = 20.0$, $L = 2$, $T = 2\pi/\Omega$). A: Asymmetric, S: symmetric, SN: saddle-node, PD: period-doubling, NB: Neimark bifurcation, P-1: period-1 motion, P-2: period-2 motion

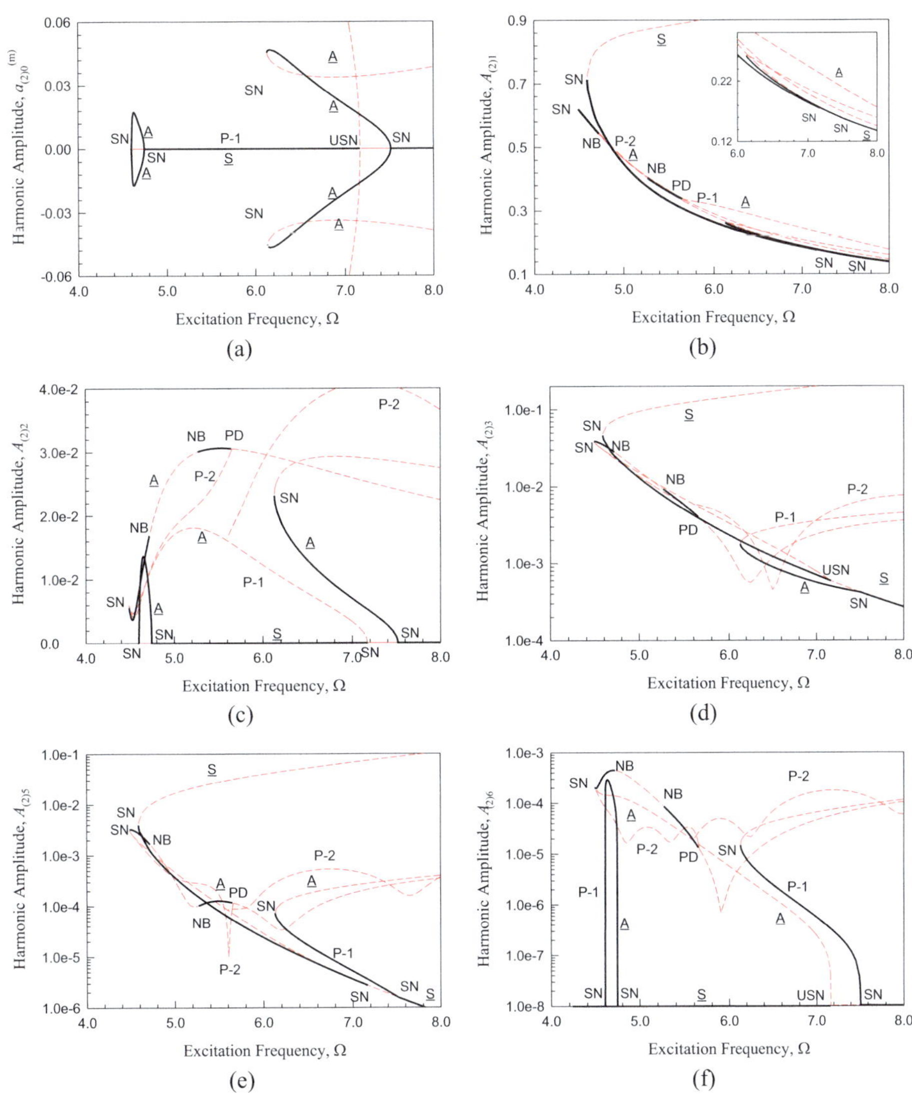

Fig. 5.19 A further zoomed view of the frequency-amplitude characteristics of bifurcation tree of period-1 to period-2 motions varying with excitation frequency of $\Omega \in (2.0, 8.0)$ for displacement x_2: **a** $a_{(2)0}$, **b–d** $A_{(2)j}$ ($j = 1, 2, 3, 5, 6$) ($m = 2$, $k_1 = 5$, $k_2 = 100$, $m_1 = 1$, $c = 0.1$, $Q_0 = 20.0$, $L = 2$, $T = 2\pi/\Omega$). A: Asymmetric, S: symmetric, SN: saddle-node, PD: period-doubling, NB: Neimark bifurcation, P-1: period-1 motion, P-2: period-2 motion

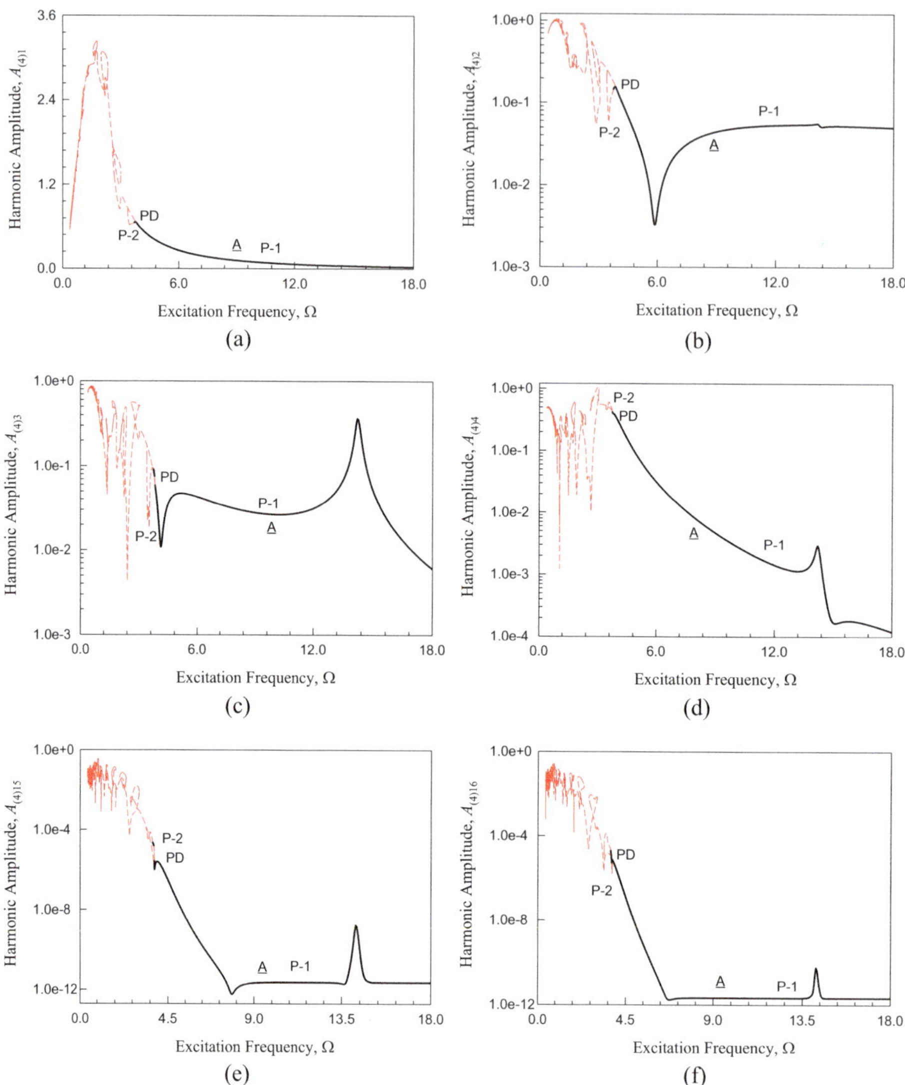

Fig. 5.20 A global view of the frequency-amplitude characteristics of bifurcation tree of period-1 to period-2 travelable motions varying with excitation frequency of $\Omega \in (0.0, 18.0)$ for velocity y_2: **a–f** $A_{(4)j}$ $(j = 1, 2, 3, 15, 16)$ $(m = 2, k_1 = 5, k_2 = 100, m_1 = 1, c = 0.1, Q_0 = 20.0, L = 2, T = 2\pi/\Omega)$. A: Asymmetric, S: symmetric, SN: saddle-node, PD: period-doubling, NB: Neimark bifurcation, P-1: period-1 motion, P-2: period-2 motion

Table 5.4 A list of zoomed views of frequency-amplitudes for displacements x_1 and x_2

	Frequency ranges Ω	Non-travelable motions	Bifurcation trees
Figures 5.4 and 5.5	(0.0, 2.25)	P-1(S) $\to$ P-1(A) $\to$ P-2 (A)	Global view
Figures 5.6 and 5.7	(0.450, 0.495)	P-1(S) $\to$ P-1(A) $\to$ P-2 (A)	1 (Stable) + 1(U) branches
Figures 5.8 and 5.9	(0.65, 0.80)	P-1(S) $\to$ P-1(A) $\to$ P-2 (A)	3 (Stable) + 1 (U) branches
Figures 5.10 and 5.11	(0.85, 1.00)	P-1(S) $\to$ P-1(A) $\to$ P-2 (A)	1 (Stable) + 2(U) branches
Figures 5.12 and 5.13	(1.05, 1.25)	P-1(S) $\to$ P-1(A) $\to$ P-2 (A)	1 (Stable) + 2(U) branches
Figures 5.13 and 5.15	(1.4, 2.0)	P-1(S) $\to$ P-1(A) $\to$ P-2 (A)	1 (Stable) branches
Figures 5.16 and 5.17	(3.6, 18.0)	P-1(S) $\to$ P-1(A) $\to$ P-2 (A)	2 (Stable) branches

Stable stable branch, *U* unstable branch. *S* symmetric period-1 motion, *A* asymmetric period-1 motion

$$x_1 = a_{(1)0} + \sum_{j=1}^{\infty} b_{(1)(2j)} \cos[(2j)\Omega t] + c_{(1)(2j)} \sin[(2j)\Omega t]$$

$$= a_{(1)0} + \sum_{j=1}^{\infty} A_{(1)(2j)} \cos[(2j)\Omega t - \varphi_{(1)(2j)}], \tag{5.2}$$

and the solution of the pendulum is

$$x_2 = \sum_{j=1}^{\infty} b_{(2)(2j-1)} \cos[(2j-1)\Omega t] + c_{(2)(2j-1)} \sin[(2j-1)\Omega t]$$

$$= \sum_{j=1}^{\infty} A_{(2)(2j-1)} \cos[(2j-1)\Omega t - \varphi_{(2)(2j-1)}]. \tag{5.3}$$

For asymmetric period-1 and period-2 motions, the analytical expressions were presented in Eqs. (5.2) and (5.3). For the non-travelable periodic motion, harmonic amplitudes decay with harmonic order increase. Thus, the harmonic amplitudes $A_{(1)15}$ and $A_{(1)16}$ for the period-1 and period-2 motions of the spring oscillator have about the quantity levels of 6×10^{-2}. The harmonic amplitudes $A_{(2)15}$ and $A_{(2)16}$ for the non-travelable period-1 and period-2 motions of the pendulum have about the quantity levels of 3×10^{-2}. For $\Omega \in (0.45, 0.495)$, there exists one stable asymmetric period-1 motion branch with bifurcation trees of period-2 motions, and there is also an unstable symmetric period-1 motion branch without bifurcation trees. The corresponding bifurcation points were listed in Tables 5.1, 5.2 and 5.3.

Similarly, the further zoomed views of frequency-amplitude characteristics for $\Omega \in (0.65, 0.80)$ are presented in Figs. 5.8 and 5.9 for the non-travelable periodic motions of the spring-pendulum systems. In Fig. 5.8a–f, the frequency-amplitude characteristics of the nonlinear spring system are presented by $a_{(1)0}^{(m)}$ and $A_{(1)(2j)/m} = A_{(1)j}$ ($j = 1, 2, 3, 15, 16$; $m = 2$) for period-1 and period-2 motions. The frequency-amplitude characteristics of the pendulum system are placed in Fig. 5.9a–f through $a_{(2)0}^{(m)}$ and $A_{(2)(2j)/m} = A_{(2)j}$ ($j = 1, 2, 3, 15, 16$; $m = 2$). For the non-travelable periodic motion, harmonic amplitudes decay with harmonic order increase. Thus, the harmonic amplitudes $A_{(1)15}$ and $A_{(1)16}$ for the period-1 and period-2 motions of the spring oscillator have about the quantity levels of 8×10^{-2}. The harmonic amplitudes $A_{(2)15}$ and $A_{(2)16}$ for the non-travelable period-1 and period-2 motions of the pendulum have about the quantity levels of 8×10^{-2}. The harmonic amplitudes of asymmetric period-1 and period-2 motions for the nonlinear spring and pendulum systems are very complicated in such a frequency range of $\Omega \in (0.65, 0.80)$. For such a frequency range, there exists three (3) stable asymmetric period-1 motion branch with bifurcation trees of period-2 motions, and there is also one (1) unstable asymmetric period-1 motion branches without bifurcation trees. The corresponding bifurcation points were listed in Tables 5.1, 5.2 and 5.3.

Similarly, the further zoomed views of frequency-amplitude characteristics for $\Omega \in (0.85, 1.00)$ are presented in Figs. 5.10 and 5.11 for the non-travelable periodic motions of the spring-pendulum systems. In Figs. 5.10a–f, the frequency-amplitude characteristics of the nonlinear spring system are presented by $a_{(1)0}^{(m)}$ and $A_{(1)(2j)/m} = A_{(1)j}$ ($j = 1, 2, 3, 15, 16$; $m = 2$) for period-1 and period-2 motions. The frequency-amplitude characteristics of the pendulum system are placed in Fig. 5.11a–f by $a_{(2)0}^{(m)}$ and $A_{(2)(2j)/m} = A_{(2)j}$ ($j = 1, 2, 3, 15, 16$; $m = 2$). The harmonic amplitudes $A_{(1)15}$ and $A_{(1)16}$ for the spring oscillator have about the quantity levels of 8×10^{-2}. The harmonic amplitudes $A_{(2)15}$ and $A_{(2)16}$ for the pendulum have about the quantity levels of 2×10^{-2}. For $\Omega \in (0.85, 1, 00)$, there exists one (1) stable asymmetric period-1 motion branch with bifurcation trees, and there are two (2) unstable asymmetric period-1 motion branches without bifurcation trees. The corresponding bifurcation points were listed in Tables 5.1, 5.2 and 5.3.

The further zoomed views of frequency-amplitude characteristics for $\Omega \in (1.05, 1.25)$ are presented in Figs. 5.12 and 5.13 for the non-travelable periodic motions of the pendulum systems. In Fig. 5.12a–f, the frequency-amplitude characteristics of the nonlinear spring system are also presented through $a_{(1)0}^{(m)}$ and $A_{(1)2j/m}$ ($j = 1, 2, 3, 15, 16$; $m = 2$). The frequency-amplitude characteristics of the pendulum system are placed in Fig. 5.13a–f through $a_{(2)0}^{(m)}$ and $A_{(2)2j/m}$ ($j = 1, 2, 3, 15, 16$;). The harmonic amplitudes $A_{(1)15}$ and

$A_{(1)16}$ for the spring oscillator have about the quantity levels of 5×10^{-2}. The harmonic amplitudes $A_{(2)15}$ and $A_{(2)16}$ for the pendulum have about the quantity levels of 5×10^{-3}. For $\Omega \in (1.05, 1.25)$, there exists one (1) stable asymmetric period-1 motion branch with bifurcation trees, and there are also two (2) unstable asymmetric period-1 motion branches without bifurcation trees. The corresponding bifurcation points were listed in Tables 5.1, 5.2 and 5.3.

The further zoomed views of frequency-amplitude characteristics for $\Omega \in (1.4, 2.0)$ are presented in Figs. 5.14 and 5.15 for the non-travelable periodic motions of the pendulum systems. In Fig. 5.14a–f, the frequency-amplitude characteristics of the nonlinear spring system are also presented through $a_{(1)0}^{(m)}$ and $A_{(1)2j/m}$ ($j = 1, 2, 3, 15, 16; m = 2$). Similarly, the symmetric period-1 motions with $A_{(1)(2l-1)} = 0$ ($l = 1, 2, \ldots$) are clearly observed. The asymmetric period-1 motions with $A_{(1)(2l-1)} \neq 0$ are obtained. The frequency-amplitude characteristics of the pendulum system are placed in Fig. 5.15a–f through $a_{(2)0}^{(m)}$ and $A_{(2)2j/m}$ ($j = 1, 2, 3, 15, 16; m = 2$). The harmonic amplitudes of asymmetric period-1 and period-2 motions for the nonlinear spring and pendulum systems are much simple in the frequency range of $\Omega \in (1.4, 2.0)$. Similarly, the symmetric period-1 motions with $a_{(2)0} = 0$ and $A_{(2)(2l)} = 0$ ($l = 1, 2, \ldots$) are clearly observed. The asymmetric period-1 motions with $a_{(2)0} \neq 0$ and $A_{(2)(2l)} \neq 0$ are obtained. The harmonic amplitudes $A_{(i)15}$ and $A_{(i)16}$ ($i = 1, 2$) for the non-travelable period-1 and period-2 motions have about the quantity levels of 10^{-3}. For $\Omega \in (1.4, 2.0)$, there exists one (1) stable asymmetric period-1 motion branch with bifurcation trees, and there are also two (2) unstable asymmetric period-1 motion branches without bifurcation trees. The corresponding bifurcation points were listed in Tables 5.1, 5.2 and 5.3.

The further zoomed views of frequency-amplitude characteristics for $\Omega \in (3.6, 18.0)$ are presented in Figs. 5.16 and 5.17 for the non-travelable periodic motions of the pendulum systems. In Fig. 5.16a–f, the frequency-amplitude characteristics of the nonlinear spring system are also presented through $a_{(1)0}^{(m)}$ and $A_{(1)2j/m}$ ($j = 1, 2, 3, 15, 16; m = 2$). The frequency-amplitude characteristics of the pendulum system are placed in Fig. 5.17a–f through $a_{(2)0}^{(m)}$ and $A_{(2)2j/m}$ ($j = 1, 2, 3, 15, 16; m = 2$). The harmonic amplitudes of asymmetric period-1 and period-2 motions for the nonlinear spring and pendulum systems are much simple in the frequency range of $\Omega \in (1.4, 2.0)$. The harmonic amplitudes $A_{(i)15}$ and $A_{(i)16}$ ($i = 1, 2$) for the non-travelable period-1 and period-2 motions have about the quantity levels of 10^{-4}. For $\Omega \in (3.6, 18.0)$, there exist two (2) stable asymmetric period-1 motion branch with bifurcation trees. The corresponding bifurcation points were listed in Tables 5.1, 5.2 and 5.3. For $\Omega \in (4.0, 8.0)$, the asymmetric period-1 and period-2 motions on the bifurcation trees are complicated and crowded. Thus, the further zoomed views of the frequency-amplitudes are presented in Figs. 5.18 and 5.19 for the spring and pendulum oscillators in the spring-pendulum system, and the bifurcation trees are clearly presented.

For the travelable motions of pendulum oscillator in the spring-pendulum system, the frequency-amplitude characteristics cannot be computed through the displacement x_2, as presented in Figs. 5.3 and 5.5. Although the finite Fourier series of pendulum displacement x_2 for the travelable motion can be carried out, the harmonic amplitudes cannot be converged to zero. Thus, the frequency-amplitude of velocity $y_2 = \dot{x}_2$ are presented through finite Fourier series analysis. For the velocity of $y_2 = \dot{x}_2$, the constant term does not exist. The harmonic amplitudes of $A_{(4)j}$ ($j = 1, 2, 3, 4, 15, 16$) are presented in Fig. 5.20 for period-1 and period-2 motions in the bifurcation trees. For the large frequency of $\Omega > 9.0$, the quantity levels for the harmonic amplitudes of $A_{(4)15}$ and $A_{(4)16}$ are close to 10^{-11}, which is totally different from the harmonic amplitudes based on the pendulum.

As previously illustrated, the harmonic amplitudes of $A_{(1)j}$ and $A_{(2)j}$ ($j = 1, 2, \ldots$) are for period-1 and period-2 motions on the bifurcation trees. However, the harmonic amplitudes of $A_{(1)(2j-1)/2}$ and $A_{(2)(2j-1)/2}$ ($j = 1, 2, \ldots$) are only for the period-2 motions of the spring oscillator and pendulum oscillator. For period-1 motions, $A_{(1)(2j-1)/2} = 0$ and $A_{(2)(2j-1)/2} = 0$ ($j = 1, 2, \ldots$). For the frequency-amplitude characteristics of period-2 motions, the harmonic amplitudes of $A_{(1)(2j-1)/2}$ and $A_{(2)(2j-1)/2}$ for $j = 1, 2, 3, 4$ are presented. The global views for the frequency-amplitude characteristics of period-2 motions are presented in Figs. 5.21 and 5.22. For lower frequency, such frequency-amplitude characteristics of period-2 motions cannot be clearly presented. Thus, the detailed views of period-2 motions should be presented in Figs. 5.23, 5.24, 5.25, 5.26, 5.27 and 5.28. For the travelable period-2 motions, the corresponding half-frequency-amplitudes of velocity y_2 are presented in Fig. 5.29. Thus, a list of the detailed zoomed views is tabulated in Table 5.5.

In Figs. 5.23 and 5.24, the half-frequency-amplitude characteristics of period-2 motions in the range of $\Omega \in (0.4, 1.0)$ are presented for displacements x_1 and x_2, respectively. There are six (6) branches of period-2 motions. The half-frequency-amplitude curves for $A_{(i)(2l-1)/2}$ ($i = 1, 2;\ l = 1, 2, 3, 4$) change for different harmonic order, and the corresponding quantity levels are almost same with 10^{-2}. In Figs. 5.25 and 5.26, the half-frequency-amplitude characteristics of period-2 motions in the range of $\Omega \in (1.0, 1.2)$ are presented for displacements x_1 and x_2, respectively. There are six (4)

Table 5.5 A list of zoomed views of frequency-amplitudes only for period-2 motions of displacements x_1 and x_2

	Frequency ranges Ω	Non-travelable motions	Bifurcation trees
Figures 5.21 and 5.22	(0.0, 18.0)	P-2 (A)	Global view
Figures 5.23 and 5.24	(0.4, 1.0)	P-2 (A)	6 branches
Figures 5.25 and 5.26	(1.0, 1.2)	P-2 (A)	4 branches
Figures 5.27 and 5.28	(1.3, 18.0)	P-2 (A)	3 branches
Figure 5.29	(3.36, 3.90)	P-2 motion	1 travelable P-2 branch

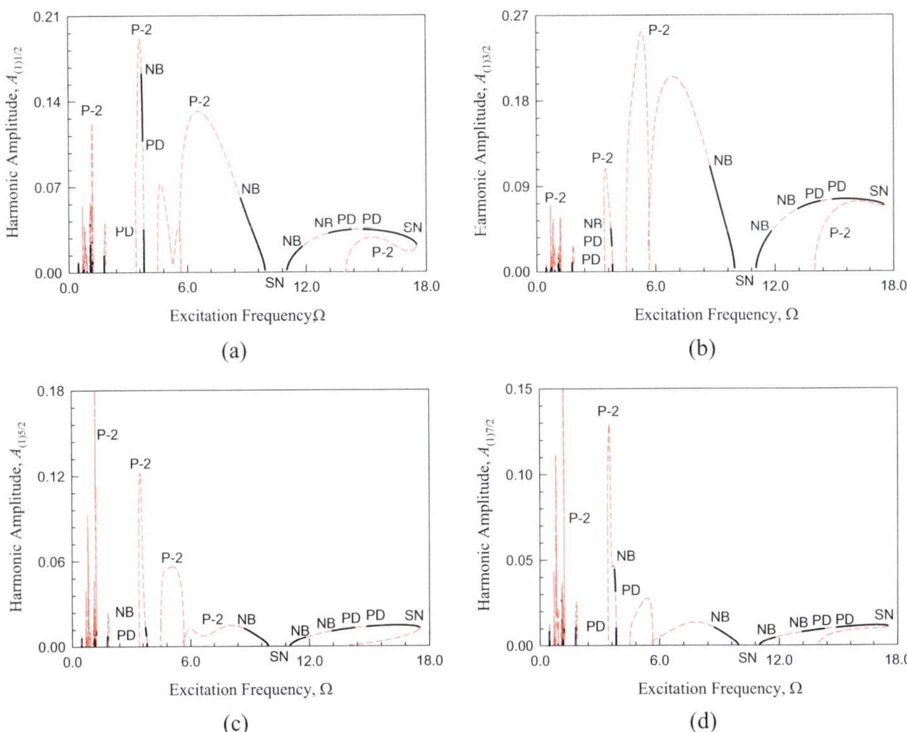

Fig. 5.21 A global view of the frequency-amplitude characteristics of bifurcation tree of period-1 to period-2 motions varying with excitation frequency of $\Omega \in (0.0, 18.0)$ for displacement x_1: **a–d** $A_{(1)k/2}$ $(k = 1, 3, 5, 7)$ $(m = 2, k_1 = 5, k_2 = 100, m_1 = 1, c = 0.1, Q_0 = 20.0, L = 2, T = 2\pi/\Omega)$. SN: saddle-node, PD: period-doubling, NB: Neiamrk bifurcation, P-2: period-2 motion

branches of period-2 motions, including three (3) branches with stable period-2 motions and one (1) branch with all unstable period-2 motion. The half-frequency-amplitude curves for $A_{(1)1/2}$, $A_{(1)3/2}$, $A_{(1)5/2}$, $A_{(1)7/2}$ change for different harmonic order, and the corresponding quantity levels are from 2×10^{-2} to 6×10^{-2}. In Figs. 5.27 and 5.28, the half-frequency-amplitude characteristics of period-2 motions in the range of $\Omega \in (1.3, 18.0)$ are presented for displacements x_1 and x_2, respectively. There are six (5) branches of period-2 motions, including four (4) branches with stable period-2 motions and one (1) branch with all unstable period-2 motion. The half-frequency-amplitude curves for $A_{(1)1/2}$, $A_{(1)3/2}$, $A_{(1)5/2}$, $A_{(1)7/2}$ change for different harmonic order, and the corresponding quantity levels are from 2×10^{-2} to 10^0. For travelable period-2 motions, the half-frequency-amplitude curves of velocity of the pendulum are presented in Fig. 5.29 with $\Omega \in (3.36, 3.90)$. Only one branch exists with the same quantity levels for $A_{(i)(2l-1)/2}$ $(i = 1, 2; l = 1, 2, 3, 4)$. The bifurcation points are listed in Table 5.3.

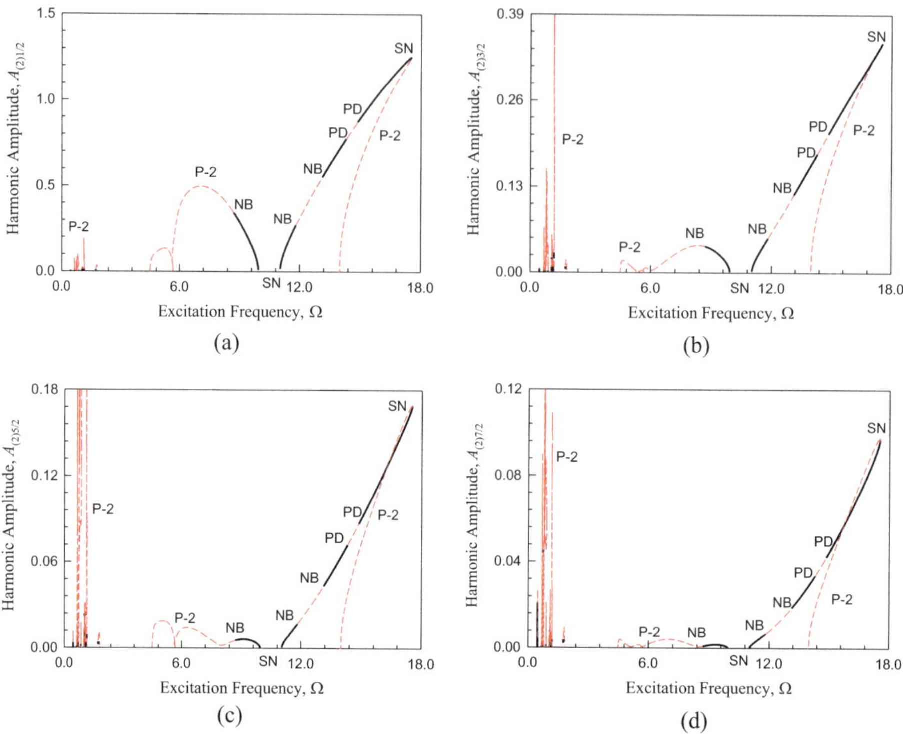

Fig. 5.22 A global view of the frequency-amplitude characteristics of bifurcation tree of period-1 to period-2 motions varying with excitation frequency of $\Omega \in (0.0, 18.0)$ for displacement x_2: **a–d** $A_{(2)k/2}$ $(k = 1, 3, 5, 7)$ $(m = 2, k_1 = 5, k_2 = 100, m_1 = 1, c = 0.1, Q_0 = 20.0, L = 2, T = 2\pi/\Omega)$. SN: saddle-node, PD: period-doubling, NB: Neiamrk bifurcation, P-2: period-2 motion

5.3   Periodic Motions Illustrations

As in Luo and Yuan [1], a numerical simulation of an asymmetric period-1 motion is presented for $\Omega = 1.76$. From the analytical prediction, the initial condition is $x_{10} \approx 0.8859$, $\dot{x}_{10} \approx -0.5554$, $x_{20} \approx 4.0603$, $\dot{x}_{20} \approx -0.4512$. The trajectory, displacement, harmonic amplitudes and phases of the spring oscillator and pendulum for such a period-1 motion in the spring pendulum system are presented in Fig. 5.30. The circular symbols are for semi-analytical results of the period-1 motion, and the solid curves are for numerical results. The blue symbols are for the initial points or periodic points to the initial conditions. The analytical and numerical solutions match very well. In Fig. 5.30i, the trajectory of the spring oscillator for the period-1 motion has six cycles, which are off the origin. The time-history of displacement for the period-1 motion of the spring oscillator are shown in Fig. 5.30ii. The numerical and analytical results also match very well. For a better understanding of the periodic motion in the spring pendulum, the harmonic spectrum of

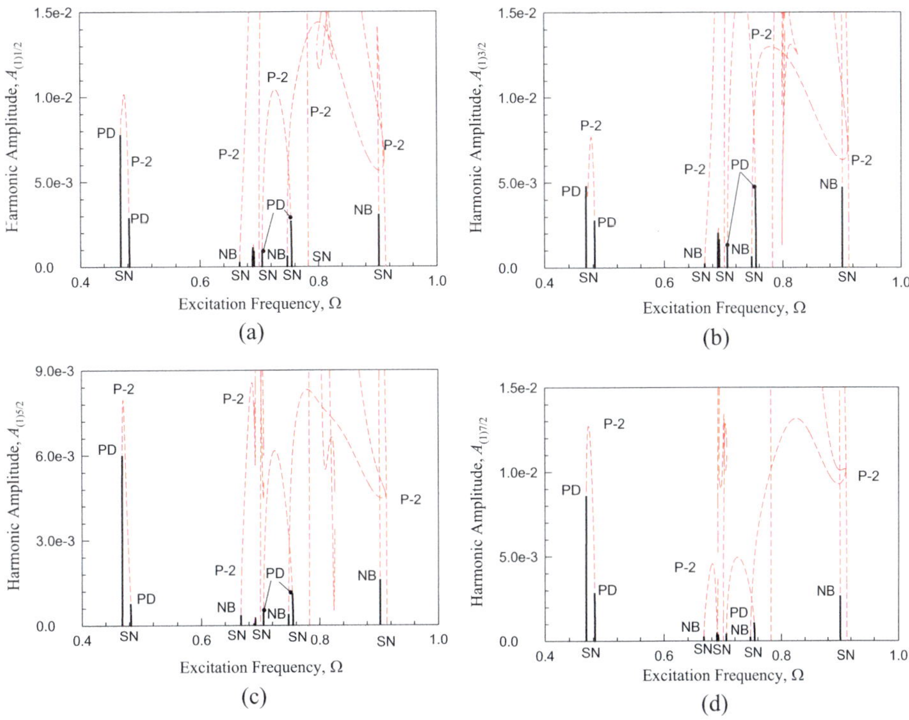

Fig. 5.23 A zoomed view of the frequency-amplitude characteristics of bifurcation tree of period-2 motions varying with excitation frequency ($\Omega \in (0.4, 1.0)$) for displacement x_1: **a–d** $A_{(1)k/2}$ ($k = 1, 3, 5, 7$) ($m = 2$, $k_1 = 5$, $k_2 = 100$, $m_1 = 1$, $c = 0.1$, $Q_0 = 20.0$, $L = 2$, $T = 2\pi/\Omega$). SN: saddle-node, PD: period-doubling, NB: Neiamrk bifurcation, P-2: period-2 motion

the spring oscillator is presented in Fig. 5.30iii. The constant term is $a_{(1)0} \approx 0.4127$. The main harmonic amplitudes are $A_{(1)1} \approx 0.0860$, $A_{(1)2} \approx 0.0815$, $A_{(1)3} \approx 0.0623$, $A_{(1)4} \approx 0.0545$, $A_{(1)5} \approx 0.1488$, $A_{(1)6} \approx 0.5099$, $A_{(1)7} \approx 0.0449$, $A_{(1)8} \approx 0.1091$, $A_{(1)9} \approx 7.7568e\text{-}3$, $A_{(1)10} \approx 7.9680e\text{-}3$, $A_{(1)11} \approx 0.0227$, $A_{(1)12} \approx 0.0411$, $A_{(1)13} \approx 7.2083e\text{-}3$, $A_{(1)14} \approx 0.0178$. Other harmonic amplitudes are $A_{(1)k} \in (10^{-11}, 10^{-2})$ ($k = 15, 16, \ldots, 80$) with $A_{(1)80} \approx 2.6133e\text{-}11$. The harmonic term of $A_{(1)6}$ plays an important role on the period-1 motion. Thus six cycles are mainly due to such a harmonic term. The corresponding harmonic phases for the period-1 motions is presented in Fig. 5.30iv. At least, the first fourteen (14) harmonic terms should be considered for the period-1 motion with accuracy of $\varepsilon = 10^{-2}$.

In Fig. 5.30v, the trajectory of pendulum for the period-1 motion in the spring-pendulum is presented, which is near zero or 2π. The pendulum is with a librational motion. The time-history of displacement for the period-1 motions of pendulum oscillator are only one sinusoidal wave, as shown in Fig. 5.30vi. The harmonic effects on periodic

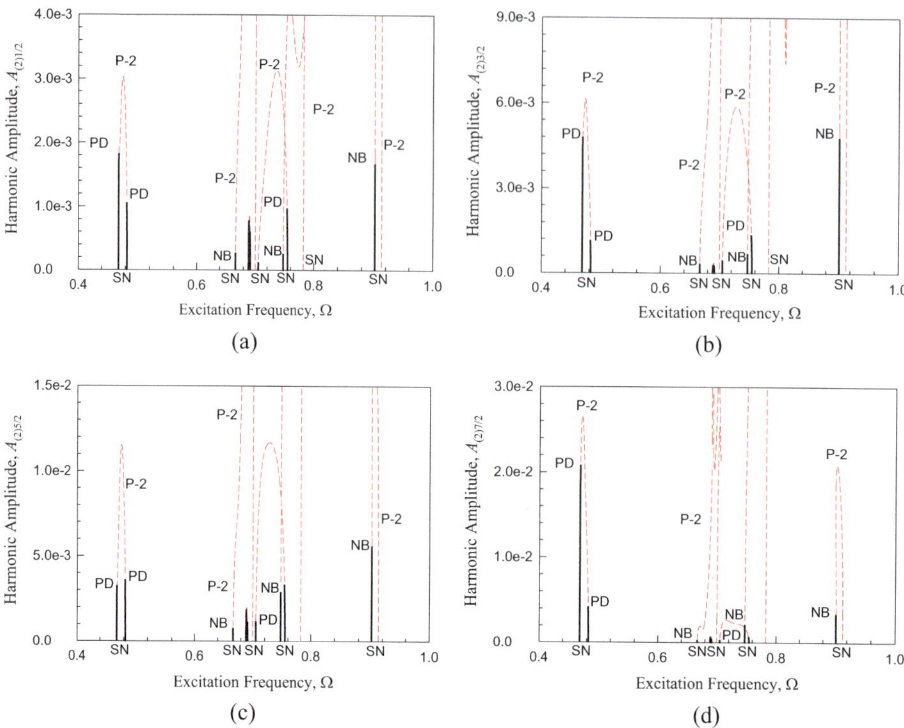

Fig. 5.24 A zoomed view of the frequency-amplitude characteristics of bifurcation tree of period-2 motions varying with excitation frequency of $\Omega \in (0.4, 1.0)$ for displacement x_2: **a–d** $A_{(2)k/2}$ $(k = 1, 3, 5, 7)$ $(m = 2, k_1 = 5, k_2 = 100, m_1 = 1, c = 0.1, Q_0 = 20.0, L = 2, T = 2\pi/\Omega)$. SN: saddle-node, PD: period-doubling, NB: Neiamrk bifurcation, P-2: period-2 motion

motions are determined by the harmonic amplitudes. In Fig. 5.30vii, the harmonic amplitudes of pendulum are presented. The constant term $a_{(2)0} \approx 2.6365\mathrm{e}{-13}$ is $A_{(2)0} = -a_{(2)0}$. The main harmonic amplitudes are $A_{(2)1} \approx 3.6699$, $A_{(2)2} \approx 0.1116$, $A_{(2)3} \approx 0.7188$, $A_{(2)4} \approx 0.1705$, $A_{(2)5} \approx 0.8223$, $A_{(2)6} \approx 0.1755$, $A_{(2)7} \approx 0.8855$, $A_{(2)8} \approx 0.1181$. Other harmonic amplitudes are $A_{(2)k} \in (10^{-11}, 10^{-2})$ $(k = 9, 10, \ldots, 80)$ with $A_{(2)80} \approx 3.1234\mathrm{e}{-11}$. The primary harmonic amplitude is one of the most important for the period-1 motions. Thus, only one cycle for the phase trajectory is observed. The harmonic phases of pendulum oscillator are presented in Fig. 5.30viii. The harmonic phases lie in the first and fourth quarters of phase. At least, the first eight (8) harmonic terms should be considered for the period-1 motion with accuracy of $\varepsilon = 10^{-2}$.

A period-2 motion in the spring pendulum is simulated for $\Omega = 1.776$. The initial conditions for the period-2 motion are $x_{10} \approx 0.8857$, $\dot{x}_{10} \approx -0.8859$, $x_{20} \approx 3.9885$ and $\dot{x}_{20} \approx -5079$. The trajectory, displacement, harmonic amplitudes and phases of such a period-2 motion of the spring and pendulum oscillators in the spring pendulum are illustrated in Fig. 5.31. In Fig. 5.31i, the trajectory of the spring oscillator

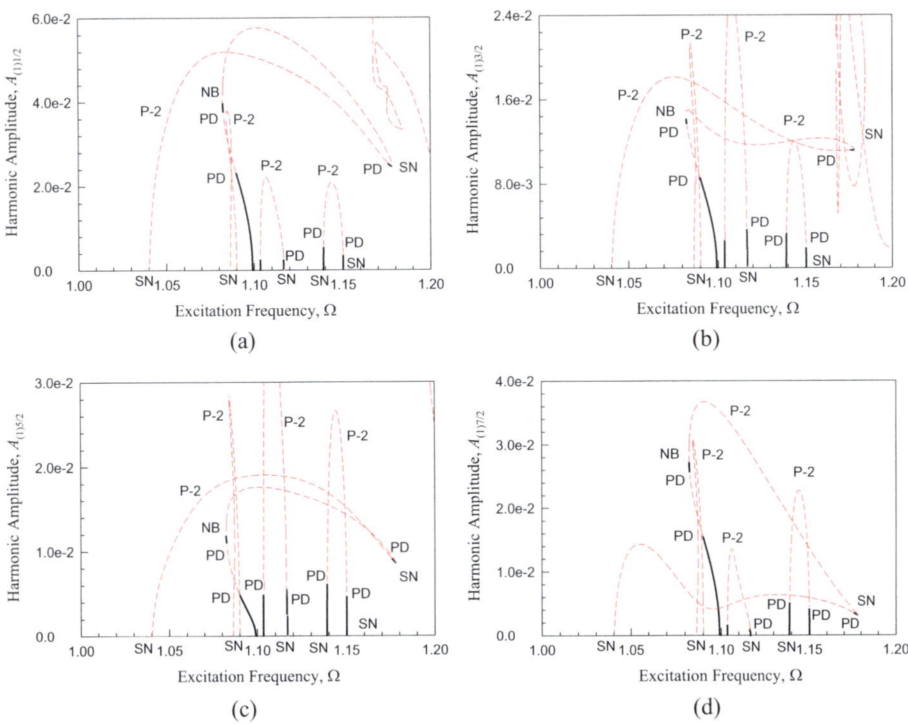

Fig. 5.25 A zoomed view of the frequency-amplitude characteristics of bifurcation tree of period-2 motions varying with excitation frequency ($\Omega \in (1.0, 1.2)$) for displacement x_1: **a–d** $A_{(1)k/2}$ ($k = 1, 3, 5, 7$) ($m = 2$, $k_1 = 5$, $k_2 = 100$, $m_1 = 1$, $c = 0.1$, $Q_0 = 20.0$, $L = 2$, $T = 2\pi/\Omega$). SN: saddle-node, PD: period-doubling, NB: Neiamrk bifurcation, P-2: period-2 motion

is presented. The trajectory of period-2 motions has doubled the trajectory of period-1 motions with twelve cycles. The corresponding time-history of displacement for the spring oscillator is presented in Fig. 5.31ii. The harmonic spectrum of the spring oscillator is presented in Fig. 5.31iii. The constant term is $a_{(1)0} \approx 0.3972$. The main harmonic terms are $A_{(1)1/2} \approx 0.0142$, $A_{(1)1} \approx 0.1293$, $A_{(1)3/2} \approx 0.0102$, $A_{(1)2} \approx 0.1025$, $A_{(1)5/2} \approx 7.9145\text{e-}3$, $A_{(1)3} \approx 0.0951$, $A_{(1)7/2} \approx 0.0110$, $A_{(1)4} \approx 0.0568$, $A_{(1)9/2} \approx 0.0176$, $A_{(1)5} \approx 0.2167$, $A_{(1)11/2} \approx 0.0217$, $A_{(1)6} \approx 0.5115$, $A_{(1)13/2} \approx 0.0160$, $A_{(1)7} \approx 0.0527$, $A_{(1)15/2} \approx 7.0320\text{e-}03$, $A_{(1)8} \approx 0.1295$, $A_{(1)17/2} \approx 4.4121\text{e-}03$, $A_{(1)9} \approx 0.0117$, $A_{(1)19/2} \approx 2.4873\text{e-}3$, $A_{(1)10} \approx 0.0114$, $A_{(1)21/2} \approx 2.7256\text{e-}3$, $A_{(1)11} \approx 0.0287$, $A_{(1)23/2} \approx 8.0148\text{e-}3$, $A_{(1)12} \approx 0.0357$, $A_{(1)25/2} \approx 2.7505\text{e-}3$, $A_{(1)13} \approx 6.8491\text{e-}3$, $A_{(1)27/2} \approx 2.1326\text{e-}3$, $A_{(1)14} \approx 0.0210$. Other Harmonic amplitudes are $A_{(1)k/2} \in (10^{-11}, 10^{-3})$ for $k = 29, 30, \ldots, 160$ with $A_{(1)80} \approx 2.8484\text{e-}12$. The harmonic phases are presented in Fig. 5.32iv. For the spring, at least, 28 harmonic terms should be considered for the accuracy of $\varepsilon = 10^{-2}$.

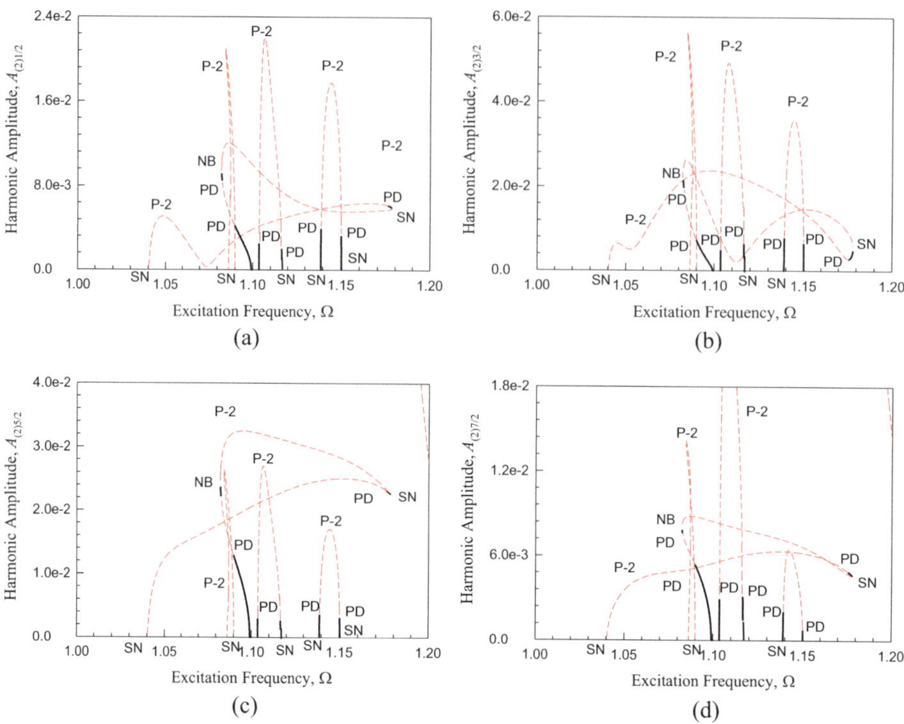

Fig. 5.26 A zoomed view of the frequency-amplitude characteristics of bifurcation tree of period-2 motions varying with excitation frequency of $\Omega \in (1.0, 1.2)$ for displacement x_2: **a–d** $A_{(2)k/2}$ ($k = 1, 3, 5, 7$) ($m = 2$, $k_1 = 5$, $k_2 = 100$, $m_1 = 1$, $c = 0.1$, $Q_0 = 20.0$, $L = 2$, $T = 2\pi/\Omega$). SN: saddle-node, PD: period-doubling, NB: Neiamrk bifurcation, P-2: period-2 motion

In Fig. 5.31v, the trajectory of the pendulum in the spring pendulum system is presented, which still is near 0 or 2π. Only two cycles for the period-2 motions are observed. The time-history of displacement of pendulum for the period-2 motion are only two waves, as shown in Fig. 5.31vi. In Fig. 5.31vii, the harmonic amplitudes of pendulum for period-2 motion is presented. The constant term is $a_{(2)0} \approx 1.9955\text{e-}03$. The main harmonic terms are $A_{(2)1/2} \approx 8.5689\text{e-}3$, $A_{(2)1} \approx 2.1498$, $A_{(2)3/2} \approx 8.5514\text{e-}3$, $A_{(2)2} \approx 0.0464$, $A_{(2)5/2} \approx 4.2683\text{e-}3$, $A_{(2)3} \approx 0.1408$, $A_{(2)7/2} \approx 3.7347\text{e-}3$, $A_{(2)4} \approx 0.0348$, $A_{(2)9/2} \approx 4.0989\text{e-}3$, $A_{(2)5} \approx 0.0981$, $A_{(2)11/2} \approx 1.0047\text{e-}3$, $A_{(2)6} \approx 0.0273$, $A_{(2)13/2} \approx 3.4685\text{e-}3$, $A_{(2)7} \approx 0.0762$, $A_{(2)15/2} \approx 2.7356\text{e-}3$, $A_{(2)8} \approx 0.0121$, $A_{(2)17/2} \approx 8.3154\text{e-}3$. $A_{(2)9} \approx 5.1998\text{e-}3$. Other harmonic amplitudes are $A_{(2)k/2} \in (10^{-11}, 10^{-3})$ for $k = 19, 20, \ldots, 160$ with $A_{(2)80} \approx 1.6915\text{e-}12$. The first harmonic amplitude is still an important harmonic term to contribute to the period-2 motions. For the period-2 motion, at least, 18 harmonic terms should be considered for the accuracy of $\varepsilon = 10^{-3}$. The corresponding harmonic phases of pendulum are presented in Fig. 5.31viii.

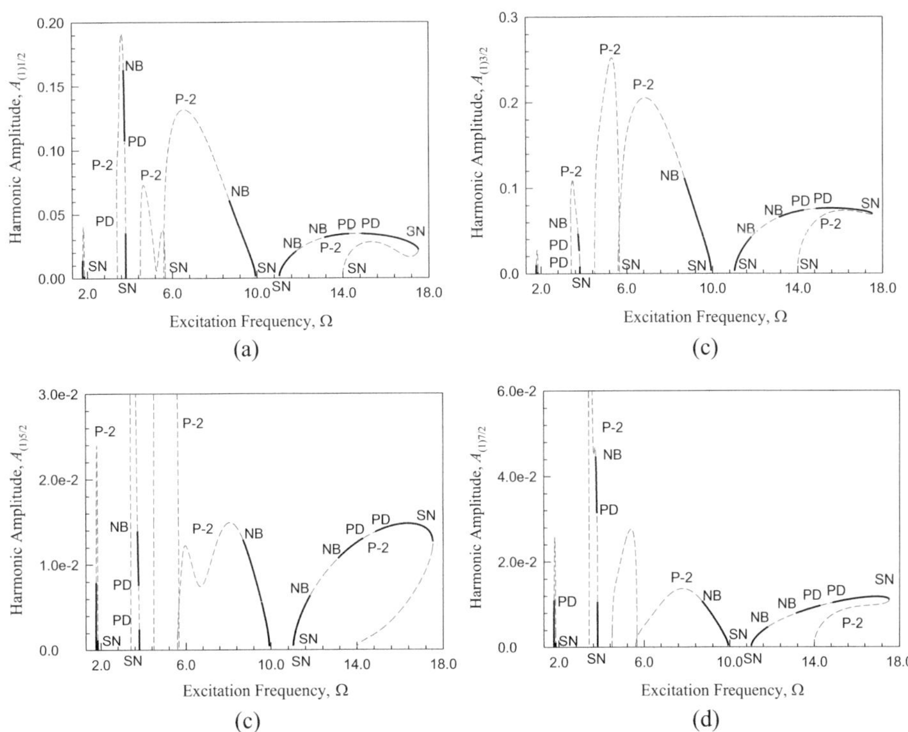

Fig. 5.27 A zoomed view of the frequency-amplitude characteristics of bifurcation tree of period-1 to period-2 motions varying with excitation frequency of $\Omega \in (1.8, 18.0)$ for displacement x_1: **a–d** $A_{(1)k/2}$ ($k = 1, 3, 5, 7$) ($m = 2$, $k_1 = 5$, $k_2 = 100$, $m_1 = 1$, $c = 0.1$, $Q_0 = 20.0$, $L = 2$, $T = 2\pi/\Omega$). SN: saddle-node, PD: period-doubling, NB: Neiamrk bifurcation, P-2: period-2 motion

A numerical simulation of a symmetric period-1 motion is presented in Fig. 5.32 for $\Omega = 8.0$. From the analytical prediction, the initial condition is $x_{10} \approx 0.4405$, $\dot{x}_{10} \approx -6.7352\text{e-}3$, $x_{20} \approx 0.1376$, $\dot{x}_{20} \approx -0.0145$. In Fig. 5.32(i), the trajectory of the spring oscillator for the period-1 motion has a simple cycle. The time-history of displacement for the period-1 motion of the spring oscillator are shown in Fig. 5.32(ii). The sinusoidal wave is presented. The harmonic amplitude spectrum of the spring oscillator is presented in Fig. 5.32(iii). The constant term is $a_{(1)0} \approx 0.4257$. The main harmonic amplitudes are $A_{(1)2} \approx 0.0148$, $A_{(1)4} = 2.9900\text{e-}6$, $A_{(1)6} \approx 7.5740\text{e-}8$, $A_{(1)8} \approx 1.1471\text{e-}10$, $A_{(1)10} \approx 7.1594\text{e-}13$, $A_{(1)12} \approx 4.4903\text{e-}14$. The harmonic term of $A_{(1)2}$ plays an important role on the period-1 motion, which is similar to the two simple sine waves. The corresponding harmonic phases for the period-1 motions is presented in Fig. 5.32(iv). The primary second order harmonic term can be considered for the period-1 motion with accuracy of $\varepsilon = 10^{-3}$.

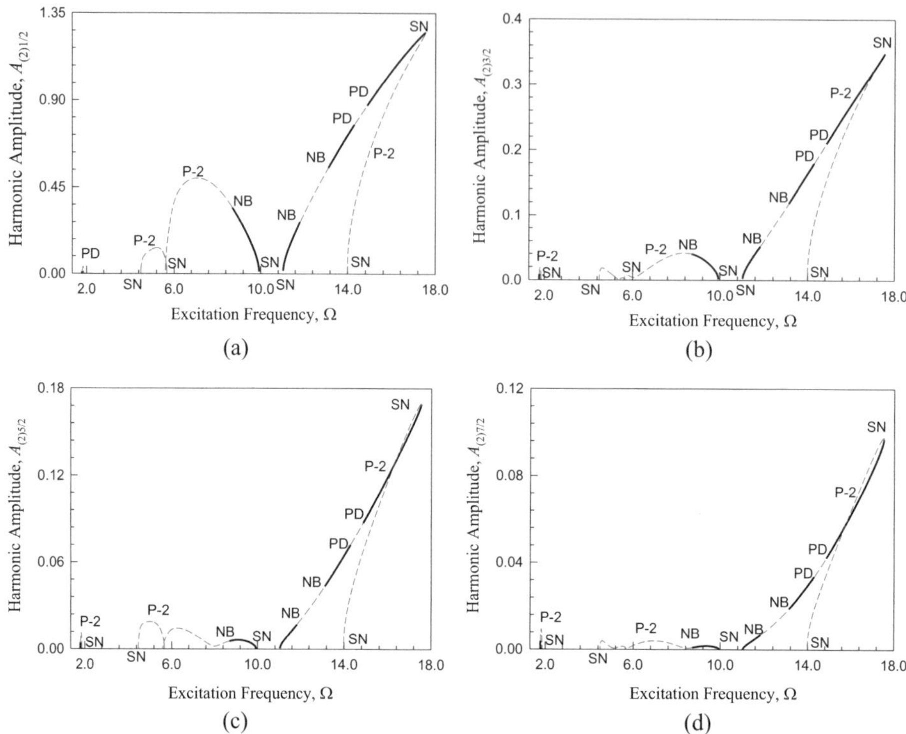

Fig. 5.28 A zoomed view of the frequency-amplitude characteristics of bifurcation tree of period-1 to period-2 motions varying with excitation frequency of $\Omega \in (2.0, 18.0)$ for displacement x_2: **a–d** $A_{(2)k/2}$ ($k = 1, 3, 5, 7$) ($m = 2$, $k_1 = 5$, $k_2 = 100$, $m_1 = 1$, $c = 0.1$, $Q_0 = 20.0$, $L = 2$, $T = 2\pi/\Omega$). SN: saddle-node, PD: period-doubling, NB: Neiamrk bifurcation, P-2: period-2 motion

$$x_1 = A_{(1)2} \cos(2\Omega t - \varphi_{(1)2}) + A_{(1)4} \cos(4\Omega t - \varphi_{(1)4}) + A_{(1)6} \cos(6\Omega t - \varphi_{(1)6}) + \cdots$$

$$= A_{(1)2} \cos(\omega_s t - \varphi_{(1)2}) + A_{(1)4} \cos(2\omega_s t - \varphi_{(1)2}) + A_{(1)5} \cos(3\omega_s t - \varphi_{(1)6}) + \cdots$$

$$\tag{5.4}$$

where ω_s is the nonlinear natural frequency of the spring system. Because the higher order harmonic terms are very small, only one harmonic term can be considered as

$$x_1 \approx A_{(1)2} \cos(2\Omega t - \varphi_{(1)2}) = A_{(1)2} \cos(\omega_s t - \varphi_{(1)2}) \tag{5.5}$$

In Fig. 5.32(v), the trajectory of pendulum for the period-1 motion in the spring-pendulum is presented, which is near zero or 2π. The time-history of displacement for the period-1 motions of pendulum oscillator are only one sine wave, as shown in Fig. 5.32(vi). The harmonic effects on periodic motions are determined by the harmonic amplitudes. In

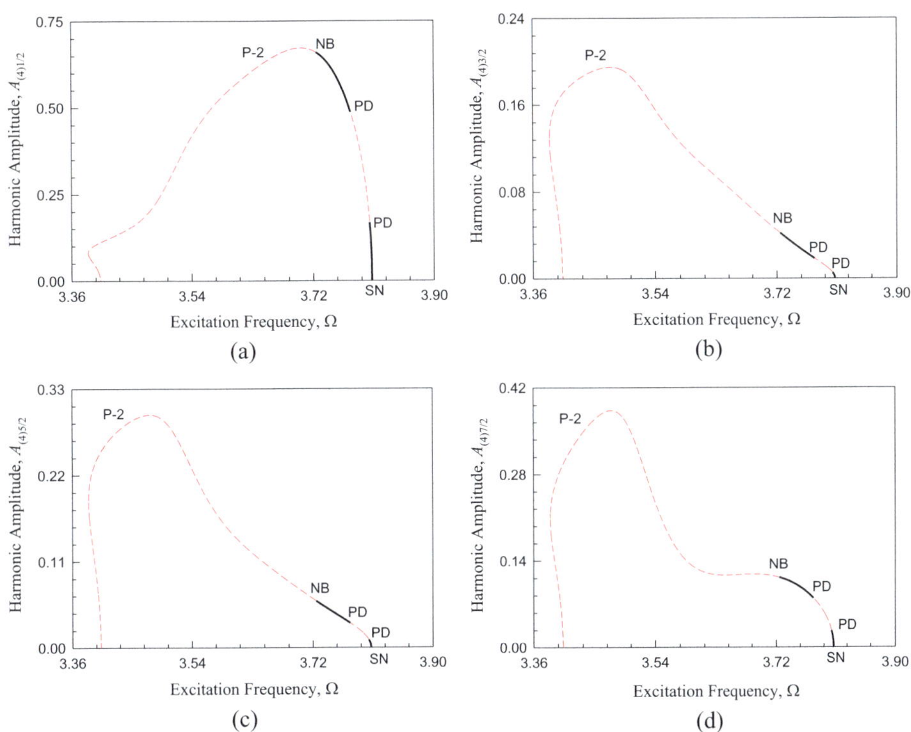

Fig. 5.29 A global view of the frequency-amplitude characteristics of bifurcation tree of period-1 to period-2 *travelable* motions varying with excitation frequency of $\Omega \in (2.0, 18.0)$ for velocity y_2: **a–d** $A_{(4)k/2}$ $(k = 1, 3, 5, 7)$ $(m = 2, k_1 = 5, k_2 = 100, m_1 = 1, c = 0.1, Q_0 = 20.0, L = 2, T = 2\pi/\Omega)$. SN: saddle-node, PD: period-doubling, NB: Neiamrk bifurcation, P-2: period-2 motion

Fig. 5.32(vii), the harmonic amplitudes of pendulum are presented. The constant term is $a_{(2)0}$. The main harmonic amplitudes are $A_{(2)1} \approx 1.1028$, $A_{(2)3} \approx 6.4624\mathrm{e}\text{-}3$, $A_{(2)5} \approx 3.1007\mathrm{e}\text{-}5$, $A_{(2)7} \approx 1.6629\mathrm{e}\text{-}7$, $A_{(2)9} \approx 8.8680\mathrm{e}\text{-}10$, $A_{(2)11} \approx 5.3835\mathrm{e}\text{-}12,.$ The primary harmonic amplitude is one of the most important for the period-1 motions. Thus, only one cycle for the phase trajectory is observed. The harmonic phases of pendulum oscillator are presented in Fig. 5.32(viii). The harmonic term of $A_{(2)1}$ plays an important role on the period-1 motion, which is similar to the simple sine wave function. The primary first order harmonic term can be considered for the period-1 motion with accuracy of $\varepsilon = 10^{-3}$.

$$x_2 = A_{(2)1}\cos(\Omega t - \varphi_{(2)1}) + A_{(2)3}\cos(3\Omega t - \varphi_{(2)3}) + A_{(2)5}\cos(5\Omega t - \varphi_{(2)5}) + \cdots$$

$$= A_{(2)1}\cos(\omega_p t - \varphi_{(2)1}) + A_{(2)3}\cos(3\omega_p t - \varphi_{(2)3}) + A_{(2)5}\cos(5\omega_p t - \varphi_{(2)5}) + \cdots$$

$$(5.6)$$

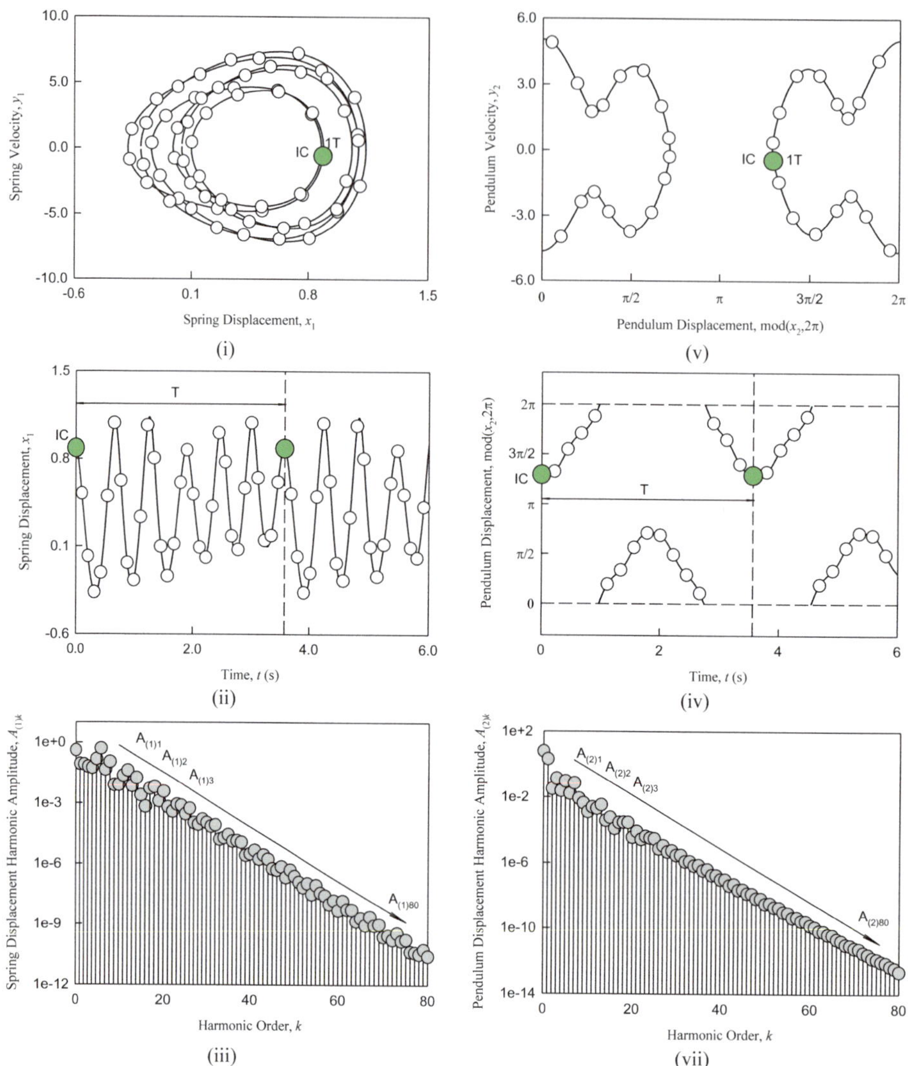

Fig. 5.30 Stable period-1 motion ($\Omega = 1.76$) for spring and pendulum systems: **i**, **v** trajectory, **ii**, **vi** displacement, **iii**, **vii** harmonic amplitudes, **iv**, **viii** harmonic phases, Parameters: $(x_{10}, \dot{x}_{10}, x_{20}, \dot{x}_{20}) \approx (0.8859, -0.5554, 4.0603, -0.4512)$. ($k_1 = 5.0, k_2 = 100.0, c = 0.1,$ $m_1 = 1.0, L = 2, Q_0 = 20.0$)

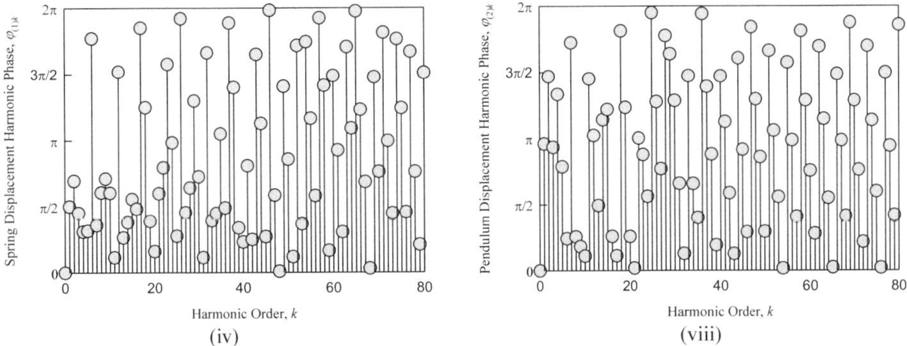

Fig. 5.30 (continued)

where ω_p is the nonlinear natural frequency of the pendulum system. Because the higher order harmonic terms are very small, only one harmonic term can be considered as

$$x_2 \approx A_{(2)1} \cos(\Omega t - \varphi_{(2)1}) = A_{(2)1} \cos(\omega_p t - \varphi_{(2)1}). \qquad (5.7)$$

Thus,

$$\omega_s = 2\omega_p, \; \varphi_{(2)1} \approx \varphi_{(1)2} \qquad (5.8)$$

because of

$$\varphi_{(1)2} = 0.02854581432547,$$
$$\varphi_{(2)1} = 0.013383412003796. \qquad (5.9)$$

Therefore,

$$x_1 \approx A_{(1)2} \cos(\omega_s t - \varphi_{(1)2}), \quad x_2 \approx A_{(2)1} \cos(\omega_p t - \varphi_{(2)1}) \qquad (5.10)$$

with $\omega_s = 2\omega_p$. Such a phenomenon can be found in the early studies [2–8].

Consider a travelable period-1 motion for $\Omega = 4.0$ with the initial condition of $x_{10} \approx 0.9467$, $\dot{x}_{10} \approx 2.0998$, $x_{20} \approx 4.4589$, $\dot{x}_{20} \approx 3.8007$. The corresponding trajectory, displacement, harmonic amplitudes and phases of the spring oscillator and pendulum are presented in Fig. 5.33. In Fig. 5.33i, the trajectory of the spring oscillator for the period-1 motion has three large cycles plus a knot. The time-history of displacement for the period-1 motion of the spring oscillator are shown in Fig. 5.33ii. The two middle waves plus a large with small wave form a periodic wave for nonlinear spring. The harmonic spectrum of the spring oscillator is presented in Fig. 5.33iii. The constant term is $a_{(1)0} \approx 0.7615$. The main harmonic amplitudes are $A_{(1)1} \approx 0.1787$, $A_{(1)2} \approx 0.1090$, $A_{(1)3} \approx 0.0124$, $A_{(1)4} \approx 0.1020$, $A_{(1)5} \approx 0.0305$. Other harmonic amplitudes are $A_{(1)k} \in (10^{-12}, 10^{-2})$

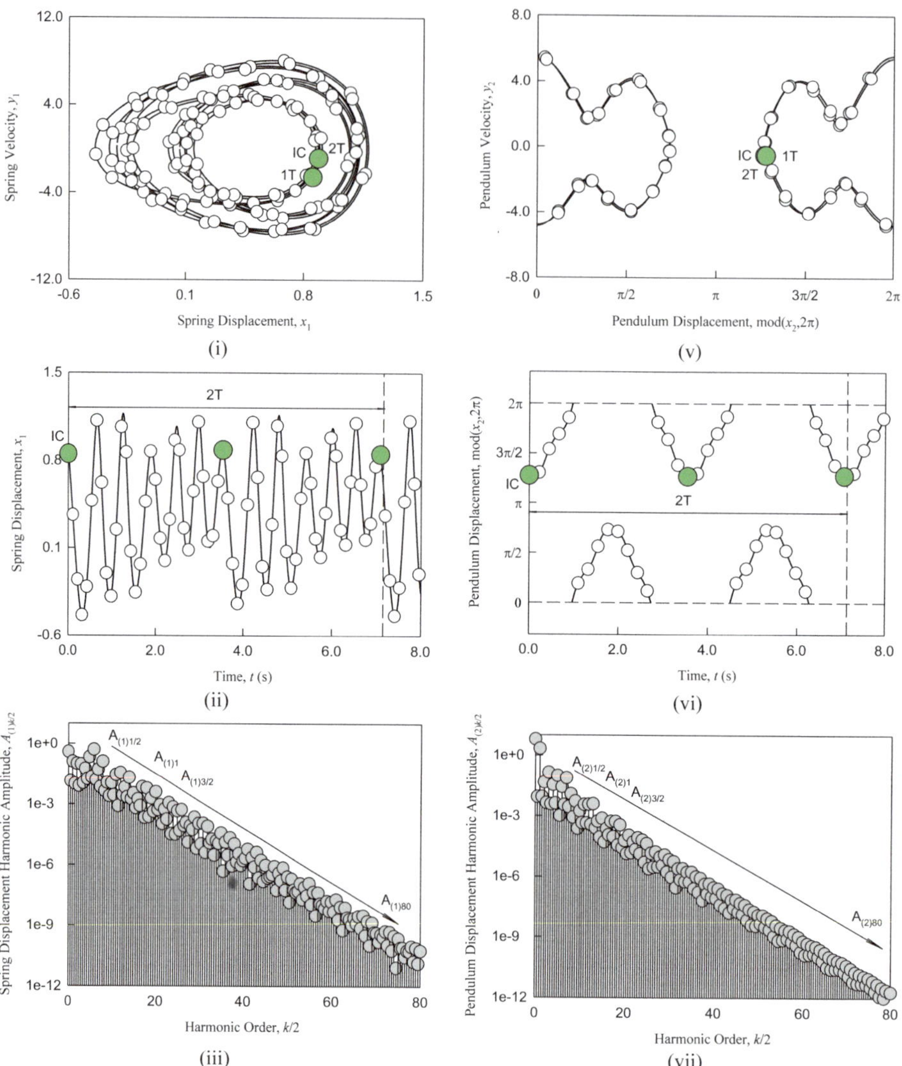

Fig. 5.31 Stable period-2 motion ($\Omega = 1.776$) for the spring and pendulum oscillators: **i**, **v** trajectory, **ii**, **vi** displacement, **iii**, **vii** harmonic amplitudes, **iv**, **viii** harmonic phases. $(x_{10}, \dot{x}_{10}, x_{20}, \dot{x}_{20}) \approx (0.8857, -0.8859, 3.9885, -0.5076)$. Parameters: $(k_1 = 5.0, k_2 = 100.0, c = 0.1, m_1 = 1.0, L = 2, Q_0 = 20.0)$

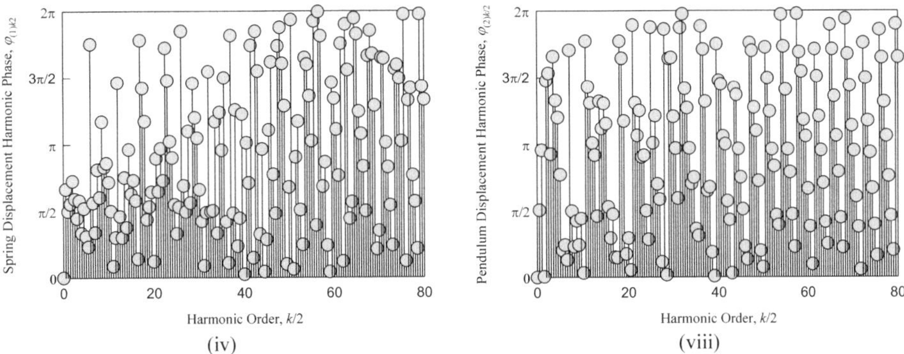

Fig. 5.31 (continued)

$(k = 6, 7, \ldots, 30)$ with $A_{(1)30} \approx 4.1306\text{e-}12$. The harmonic terms of $A_{(1)1}$ and $A_{1(2)}$ have significant contributions on the period-1 motion. Thus the two middle cycles plus a large cycle are mainly due to such a harmonic term. The corresponding harmonic phases for the period-1 motions is presented in Fig. 5.33iv. At least, the first five (5) harmonic terms should be considered for the period-1 motion with accuracy of $\varepsilon = 10^{-2}$.

In Fig. 5.33v, the trajectory of pendulum for the period-1 motion in the spring-pendulum is presented. The phase trajectory cannot make a cycle. The final and initial displacements have a difference of 2π. Thus, such a rotational motion is called travelable motion of the pendulum. The time-history of displacement for the pendulum oscillator are shown in Fig. 5.33vi. The difference between the final and initial displacements are 2π, which is clearly presented. Because the angular displacement of pendulum is closed for the travelable period-1 motions, the Fourier series cannot be applied to the angular displacement of the travelable periodic motions, Thus, the harmonic characteristics of the travelable period-1 motion will be presented through the velocity. In Fig. 5.33vii, the harmonic amplitudes of velocity for the travelable period-1 motions of pendulum are presented. The constant term $a_{(4)0} = 4.0$. The main harmonic amplitudes are $A_{(4)1} \approx 0.5927$, $A_{(4)2} \approx 0.1242$, $A_{(4)3} \approx 0.0189$, $A_{(4)4} \approx 0.3058$, $A_{(4)5} \approx 0.0979$. Other harmonic amplitudes are $A_{(4)k} \in (10^{-12}, 10^{-2})$ $(k = 6, 7, \ldots, 30)$ with $A_{(4)30} \approx 9.6186\text{e-}12$. The first primary harmonic amplitude is one of the most important for such a travelable period-1 motions. The harmonic phases of pendulum oscillator are presented in Fig. 5.33viii. The first five (5) harmonic terms should be considered for the travelable period-1 motion with accuracy of $\varepsilon = 10^{-2}$.

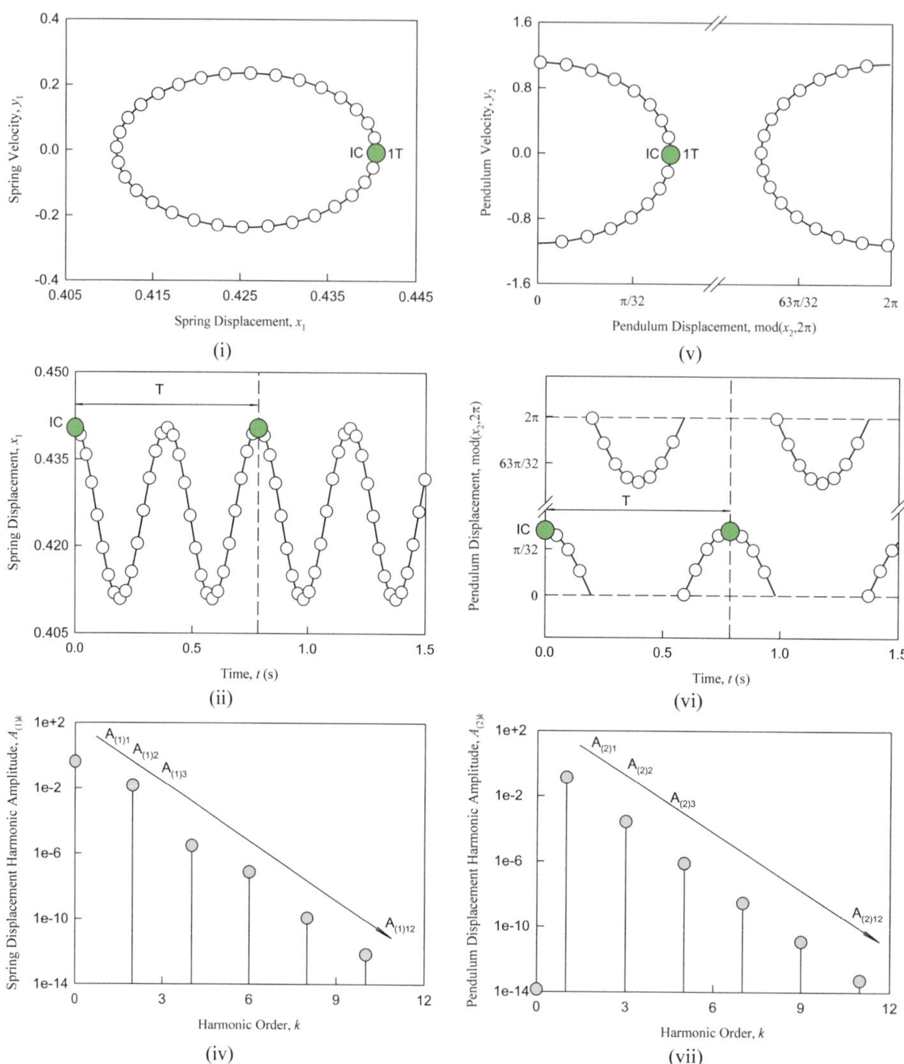

Fig. 5.32 Stable period-1 motion ($\Omega = 8.0$) for the spring and pendulum oscillators: **i, v** trajectory, **ii, vi** displacement, **iii, vii** harmonic amplitudes, **iv, viii** harmonic phases. ($x_{10}, \dot{x}_{10}, x_{20}, \dot{x}_{20}) \approx$ (0.4405, −6.7275e−3, 0.1376, −0.0145). Parameters: ($k_1 = 5.0, k_2 = 100.0, c = 0.1, m_1 = 1.0, L = 2, Q_0 = 20.0$)

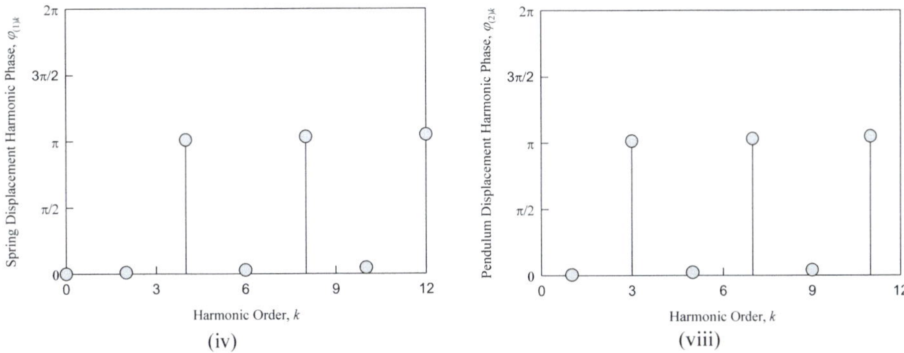

Fig. 5.32 (continued)

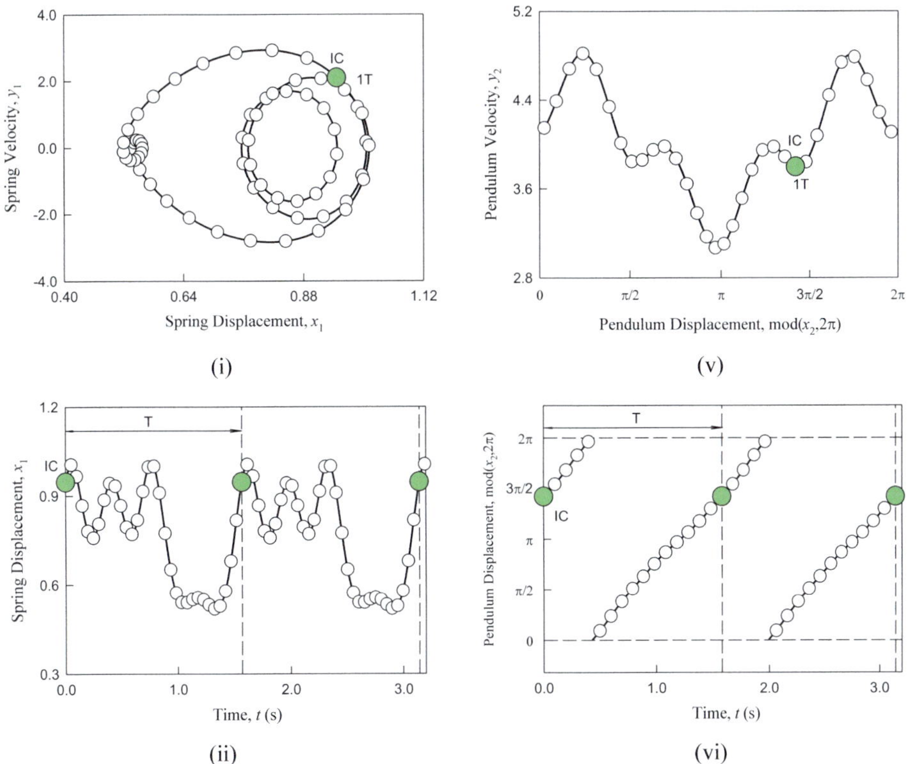

Fig. 5.33 A stable travelable period-1 motion ($\Omega = 4.0$) for the spring and pendulum oscillators: **i, v** trajectory, **ii, vi** displacement, **iii, vii** harmonic amplitudes, **iv, viii** harmonic phases. $(x_{10}, \dot{x}_{10}, x_{20}, \dot{x}_{20}) \approx (0.9467, 2.0998, 4.4589, 3.8007)$. Parameters: $(k_1 = 5.0, k_2 = 100.0, c = 0.1, m_1 = 1.0, L = 2, Q_0 = 20.0)$

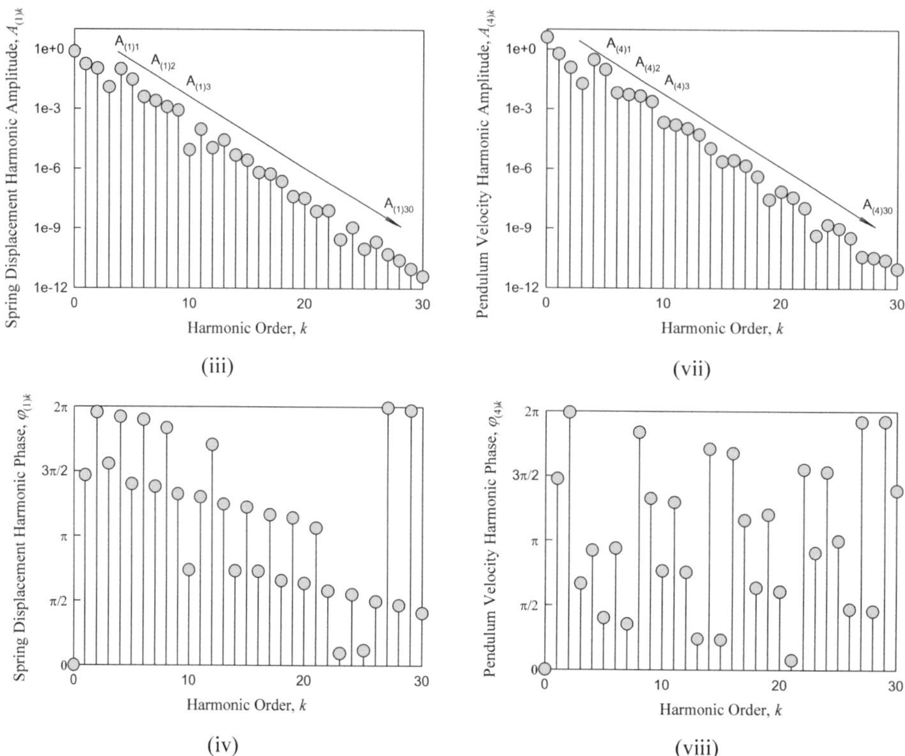

Fig. 5.33 (continued)

References

1. Guo Y (2022) Bifurcations and harmonic responses of period-1 motions in a periodically excited spring pendulum. *Journal of Vibration Testing and System Dynamics*, **6**(3):297–315.
2. Witt A, Gorelik G (1933) Kolebanya uprugogo mayatnika, kak primer kolebanii dvuh parametricheski svyazannykh lineinikh sistem, *Zh. Tekh. Fiz.*, **3**:294–306. (The oscillation of an elastic pendulum as an example of the oscillations of two parametrically connected linear systems. *Zh. Tekh. Fiz.*, **3**(2–3):294–306.)
3. Olsson MG (1976) Why does amass on a spring sometimes misbehave? *American Journal of Physics*, **44**(12):1211–1212.
4. Falk L(1978) Recurrence effects in the parametric spring pendulum. *American Journal of Physics* **46**(11):120–1123.
5. Lai HM (1984) On the recurrence phenomenon of a resonant spring pendulum. *American Journal of Physics*, **52**(3): 219–223.
6. Nunez-Yepez HN, Salas-Brito AL, Vargas CA, Vicente L (1990) Onset of chaos in an extensible pendulum, *Physics Letters A*, **145**:101–105.

7. Cuerno R, Ranada AF, Ruiz-Lorenzo JJ (1992) Deterministic chaos in the elastic pendulum: A simple laboratory for non-linear dynamics. *American Journal of Physics*, **60**(1): 73–79.
8. Bayly PV, Virgin LN (1993) An empirical investigation of the stability of periodic motion in the forced spring-pendulum. *Proceedings of the Royal Society of London*, **443A**:391–408.

6.1 Analytical Bifurcation Trees

The complete analytical bifurcation trees varying excitation amplitude Q_0 will be presented for period-3 motions to chaos. The stability and bifurcations of periodic motions will be determined through eigenvalues. Further, the bifurcation trees of the corresponding harmonic amplitudes for such periodic motions will also be presented.

As in Guo and Luo [1], consider a set of system parameters as

$$\alpha = 10.0, \quad \beta = 20.0, \quad \delta = 0.6, \quad L = 2.0, \quad \Omega = 2.0. \tag{6.1}$$

There are two branches of co-existing bifurcation trees for such period-3 motions. The first branch (B1) is presented in Fig. 6.1 with the zoomed views in Figs. 6.2, 6.3 and 6.4, while the other branch (B2) is illustrated in Fig. 6.5 with the zoomed views in Figs. 6.6, 6.7, 6.8 and 6.9. In such illustrations, the solid and dashed curves represent stable and unstable motions, respectively. Black and red colors indicate pairs of coexisting asymmetric motions, respectively. The acronyms 'SN', 'PD', 'NM', and 'USN' stand for the saddle-nodes, period-doubling, Neimark, and unstable saddle-node bifurcations, respectively. The bifurcation points are tabulated in Table 6.1. In Figs. 6.1a, b and 6.5a, b, the predictions of displacement and velocity for the spring motions are presented, represented by periodic nodes $x_{1(\mathrm{mod}(k,N))}$ and $y_{1(\mathrm{mod}(k,N))}$ for $\mathrm{mod}(k, N) = 0$, respectively. The predictions of displacement and velocity for the pendulum motions are demonstrated in Fig. 6.1c, d and 6.5c, d through the periodic nodes $\mathrm{mod}(x_{2(\mathrm{mod}(k,N))}, 2\pi)$ and $y_{2(\mathrm{mod}(k,N))}$ for $\mathrm{mod}(k, N) = 0$, respectively. The symmetric and asymmetric periodic motions are indicated by "S" and "A", respectively. In general, a period-m motion is labeled as 'P-m', with $m = 1, 2, 3, \ldots$ (i.e., period-1 motions marked by 'P-1'). The real parts and magnitudes of eigenvalues are in n Figs. 6.4e, f and 6.8e, f, respectively. Such eigenvalues are

© The Author(s), under exclusive license to Springer Nature Switzerland AG 2023   79
Y. Guo and A. C. J. Luo, *Periodic Motions to Chaos in a Spring-Pendulum System*,
Synthesis Lectures on Mechanical Engineering,
https://doi.org/10.1007/978-3-031-17883-2_6

analyzed for stability and bifurcation bifurcations of periodic motions. In such bifurcation trees, the saddle-node bifurcations of the symmetric period-3 motions lead to a pair of asymmetric period-3 motions. The symmetric period-3 motions become unstable after the bifurcations. When the asymmetric period-3 motions have period-doubling bifurcations, they become unstable and a pair of asymmetric period-6 motions will be introduced. Possible cascaded period-doubling could happen thereafter leading to period-12, period-24 ... motions. The ranges of such stable motions would shrink rapidly as the cascaded period doubling develop, which would lead to possible chaos. On the other hand, when the Neimark bifurcation happens, it would also cause the motions to turn unstable. However, no new motions are introduced. Finally, when a saddle node bifurcation associated with the jumping phenomena happens, the motion is from stable to unstable. Once the unstable saddle node bifurcations associated with jumping phenomenon occur, the corresponding periodic motions are from unstable to unstable.

In Fig. 6.2, the bifurcation trees (B1) of period-3 to period-6 motions are presented for the range of $Q_0 \in (10, 40)$. In order to show more details, zoomed views of such bifurcation trees are presented in Figs. 6.2 and 6.3 for $Q_0 \in (12, 15)$ and $(17, 23)$, respectively. In Fig. 6.4, the zoomed detailed views of the ranges of $Q_0 \in (28, 34)$, $(28.85, 29.00)$, and $(31.5, 33.5)$ are illustrated for the pendulum displacement at periodic nodes $\mod(x_{2(\mod(k,N))}, 2\pi)$ only. The bifurcation trees (B1) start with the symmetric period-3 motions. Starting from $Q_0 = 12.5844$, the symmetric period-3 motion is stable for $Q_0 \in (12.5844, 13.4250)$, $(22.1400, 22.5100)$, and $(28.9022, 28.9100)$. All three stable ranges are connected by unstable symmetric period-3 motions. At $Q_0 \approx 12.5844$, a saddle-node bifurcation associated with the jumping phenomena occurs for such symmetric period-3 motions being from stable to unstable. The symmetric period-3 motion becomes unstable and no asymmetric motions are introduced. From $Q_0 \approx 13.4250$ to 22.1400, the unstable motions are enclosed by a pair of saddle-node bifurcations existing at $Q_0 \approx 13.4250, 22.1400$. Such saddle-node bifurcations are also associated with the onset of asymmetric motions. From $Q_0 \approx 22.5100$ to 28.9022, the unstable symmetric period-3 motion is enclosed by a Neimark bifurcation and a saddle-node bifurcation. At $Q_0 \approx 22.5100$, the Neimark bifurcation is from stable to unstable while no new motions are introduced. At $Q_0 \approx 28.9022$, the saddle-node bifurcation is from unstable to stable and a pair of unstable asymmetric period-3 motions are introduced. Details of such saddle-node bifurcation at $Q_0 \approx 28.9022$ are illustrated in the zoomed view in Fig. 6.4b. Finally, at $Q_0 \approx 28.9100$, another Neimark bifurcation happens, where the symmetric period-3 motion switches from stable to unstable.

From Fig. 6.1, three branches of asymmetric period-3 motions are observed. Such coexisting pair of asymmetric period-3 motions are introduced through the saddle-node bifurcations of symmetric period-3 motions. The first branch of asymmetric period-3 motions take place from the saddle-node bifurcation at $Q_0 \approx 13.4250$. They are stable for the range of $Q_0 \in (13.4250, 13.6787)$. At $Q_0 \approx 13.6787$, the period-doubling bifurcations take place, where the asymmetric period-3 motions switch from stable to unstable.

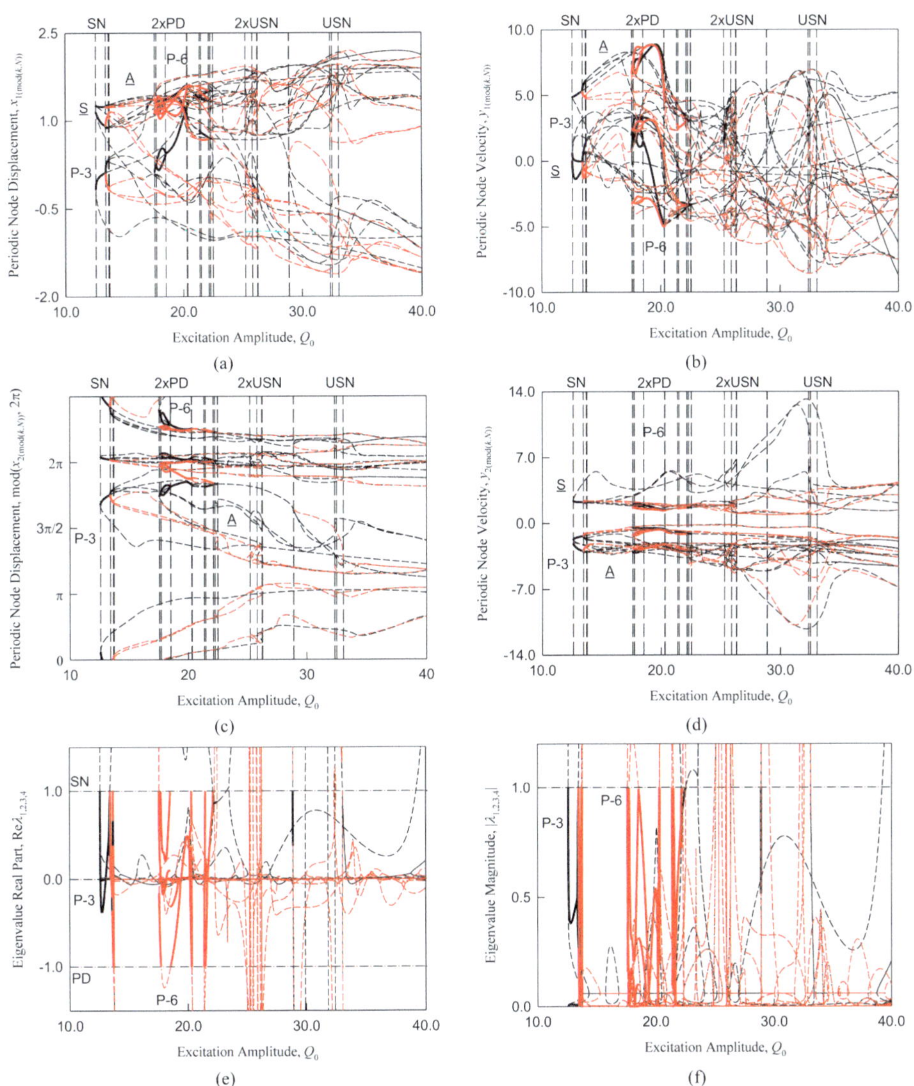

Fig. 6.1 Bifurcation tree of period-3 to chaos varying with excitation amplitude Q_0 (B1). **a** Spring node displacement $x_{1(\mathrm{mod}(k,N))}$, **b** spring node velocity $y_{1(\mathrm{mod}(k,N))}$, **c** pendulum node displacement $\mathrm{mod}(x_{2(\mathrm{mod}(k,N))}, 2\pi)$, **d** pendulum node velocity $y_{2(\mathrm{mod}(k,N))}$, **e** real part of eigenvalues, **f** eigenvalue magnitudes. ($\alpha = 10.0$, $\beta = 20.0$, $\delta = 0.6$, $L = 2.0$, $\Omega = 2.0$). SN: saddle-node, PD: period-doubling, USN: unstable saddle-node (saddle-unstable node), P-3: period-3 motion, P-6: period-6 motion

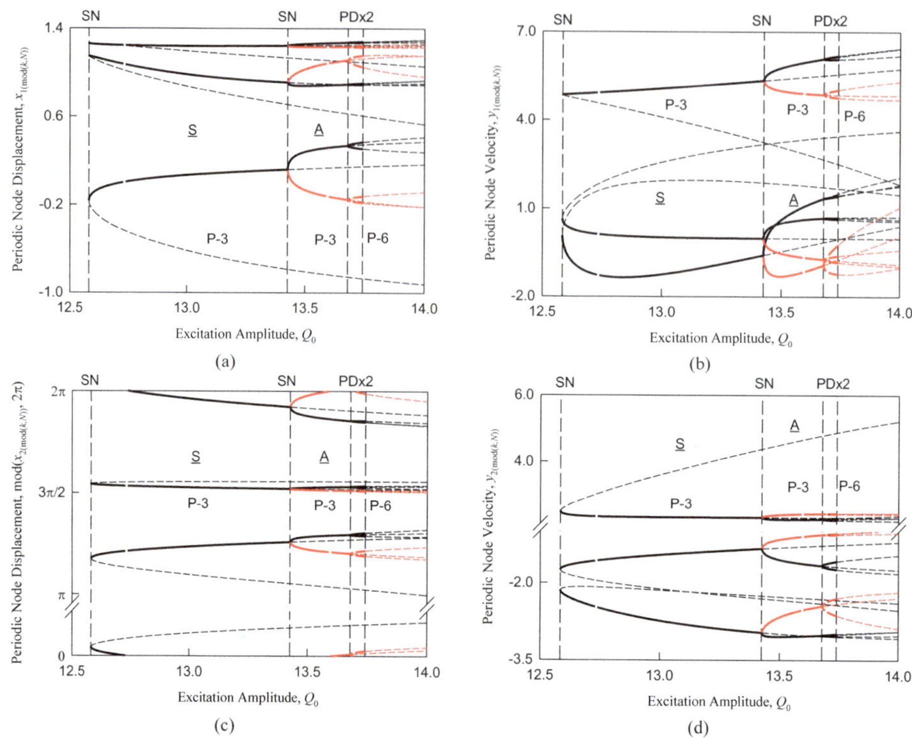

Fig. 6.2 Zoomed view of B1 for $Q_0 \in (12.0, \ 15.0)$. **a** Spring node displacement $x_{1(\mathrm{mod}(k,N))}$, **b** spring node velocity $y_{1(\mathrm{mod}(k,N))}$, **c** pendulum node displacement $\mathrm{mod}(x_{2(\mathrm{mod}(k,N))}, 2\pi)$, **d** pendulum node velocity $y_{2(\mathrm{mod}(k,N))}$ ($\alpha = 10.0$, $\delta = 0.6$, $L = 2.0$, $\Omega = 2.0$). SN: saddle-node, PD: period-doubling, USN: unstable saddle-node (saddle-unstable node), P-3: period-3 motion, P-6: period-6 motion

Such period-doubling bifurcations also introduce period-6 motions and further cascaded period-doublings. The unstable asymmetric period-3 motions encounter two pairs of unstable-saddle-node bifurcations associated with jumping phenomenon from unstable to unstable. The first pair of unstable-saddle-node bifurcation exist at $Q_0 \approx 22.2753$ and 22.5087. As the excitation amplitude Q_0 increases, the unstable asymmetric peirod-3 motions first crossover the unstable-saddle-node bifurcations at $Q_0 \approx 22.5087$. The associated jumping phenomenon causes the motions to turn towards the unstable-saddle-node bifurcations at $Q_0 \approx 22.2753$, where the jumping phenomenon turns the motion towards the second pair of the unstable-saddle-node bifurcations at $Q_0 \approx 32.3473$ and $Q_0 \approx 32.5115$. At $Q_0 \approx 32.3473$, the unstable-saddle-node bifurcation is associated with a jumping phenomenon where the unstable period-3 motion turns towards the unstable-saddle-node bifurcation at $Q_0 \approx 32.5115$. The unstable-saddle-node bifurcation at $Q_0 \approx 32.5115$ is again associated with jumping phenomenon from unstable

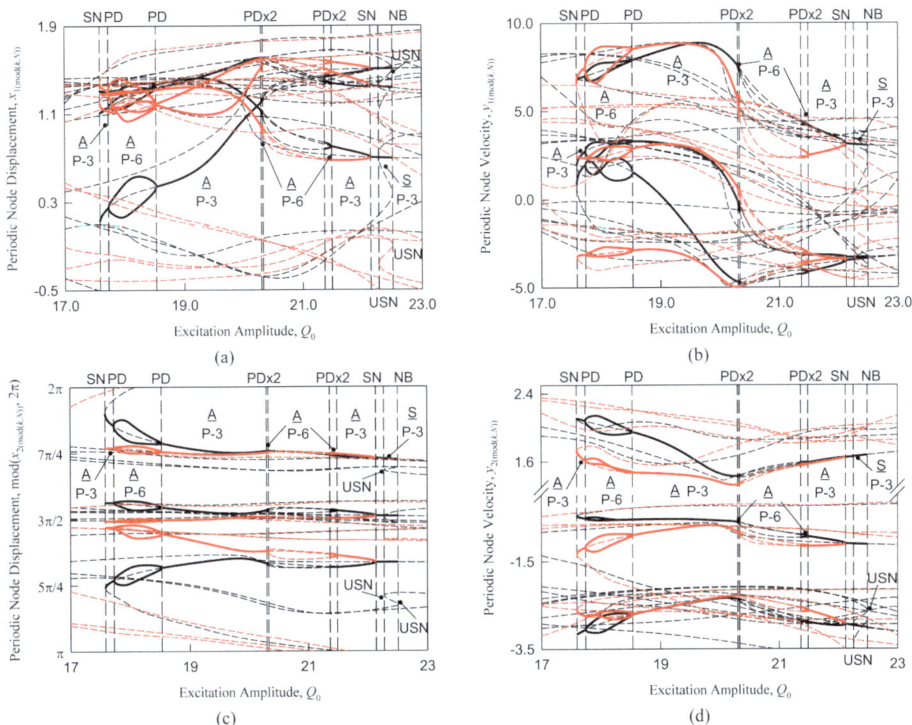

Fig. 6.3 Zoomed view of B1 for $Q_0 \in (17.0, \ 23.0)$. **a** Spring node displacement $x_{1(\mathrm{mod}(k,N))}$, **b** spring node velocity $y_{1(\mathrm{mod}(k,N))}$, **c** pendulum node displacement $\mathrm{mod}(x_{2(\mathrm{mod}(k,N))}, 2\pi)$, **d** pendulum node velocity $y_{2(\mathrm{mod}(k,N))}$ ($\alpha = 10.0, \beta = 20.0, \delta = 0.6, L = 2.0, \Omega = 2.0$). SN: saddle-node, PD: period-doubling, USN: unstable saddle-node (saddle-unstable node), P-3: period-3 motion, P-6: period-6 motion

to unstable. The second branch of asymmetric period-3 motions are stable for ranges of $Q_0 \in (21.4880, 22.1400), (18.5290, 20.3040),$ and $(17.5863, 17.7300)$. At $Q_0 \approx 22.1400$, the period-doubling bifurcations take place, where the asymmetric period-3 motions are from stable to unstable. Such period-doubling bifurcations also lead to a pair of asymmetric period-6 motions, which are connected to the period-doubling bifurcations at $Q_0 \approx 18.5290$. Similarly, another pair of asymmetric period-6 motions are introduced between the period-doubling bifurcations at. $Q_0 \approx 17.7300$ and $Q_0 \approx 17.5863$. At $Q_0 \approx 17.7300$, the saddle-node bifurcations are associated with jumping phenomenon from stable to unstable. The associated unstable asymmetric period-3 motions encounter a pair of unstable-saddle-node bifurcations at $Q_0 \approx 26.2918$ and 25.8532 as the excitation amplitude Q_0 increases. Such motions first come across the unstable-saddle-node bifurcations at $Q_0 \approx 26.2918$ where jumping phenomenon from unstable to unstable occur. The

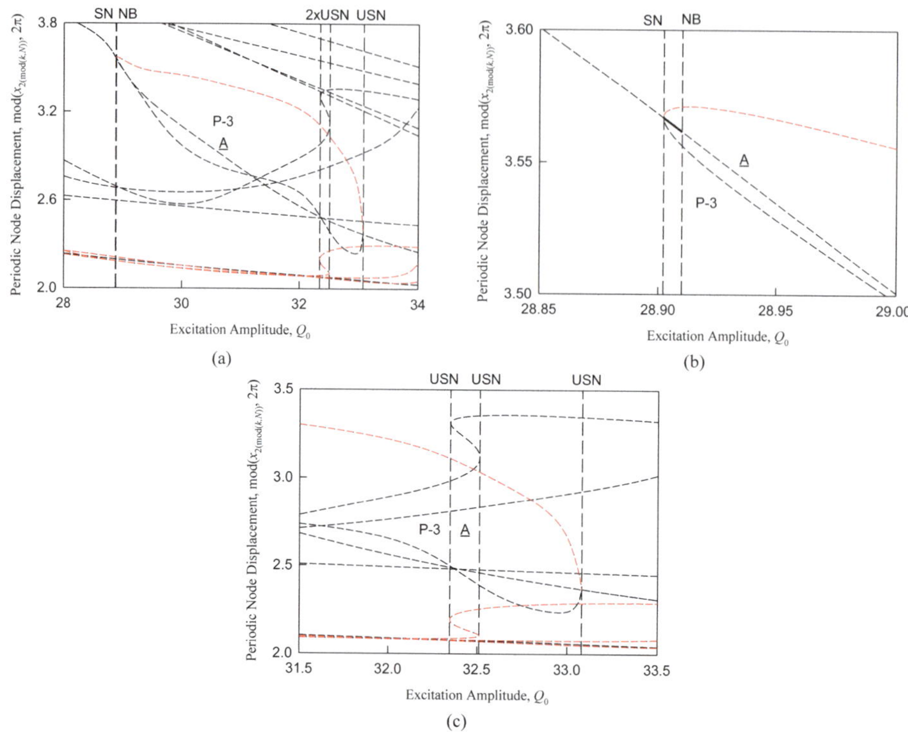

Fig. 6.4 Additional zoomed view of B1 for the periodic nodes of pendulum displacement $\mathrm{mod}(x_{2(\mathrm{mod}(k,N))}, 2\pi)$. **a** Overall view for $Q_0 \in (28.0,\ 34.0)$, **b** further zoomed view for $Q_0 \in (28.85,\ 29.00)$, **c** further zoomed view for $Q_0 \in (31.5,\ 33.5)$ ($\alpha = 10.0, \beta = 20.0, \delta = 0.6, L = 2.0, \Omega = 2.0$)

associated unstable motions continuously develop toward the unstable-saddle-node bifurcations at $Q_0 \approx 25.8532$ where a second jumping phenomenon from unstable to unstable is observed. Finally, the third branch of asymmetric period-3 motions is completely unstable. Such a pair of unstable asymmetric period-3 motions exists between the saddle-node bifurcation at $Q_0 \approx 28.9022$ and the unstable-saddle-node bifurcation at $Q_0 \approx 33.0800$. More details can be observed from the zoomed views of bifurcation trees for the periodic nodes of pendulum displacement $\mathrm{mod}(x_{2(\mathrm{mod}(k,N))},\ 2\pi)$ presented in Fig. 6.4a–c.

There are three different branches of asymmetric period-6 motions existing on bifurcation trees (B1). They are stable for the ranges of $Q_0 \in (13.678, 13.7400)$, $(17.5863, 18.5290)$, and $(20.3040, 21.4880)$. On the first branch, such asymmetric period-6 motions are stable for $Q_0 \in (13.6787, 13.7400)$. At $Q_0 \approx 13.6787$, the period-6 motions encounter saddle-node bifurcations which are associated with the period-doubling bifurcations of asymmetric period-3 motions. At $Q_0 \approx 13.7400$, period-doubling bifurcations occur, where the period-6 motions are from stable to unstable and further cascaded

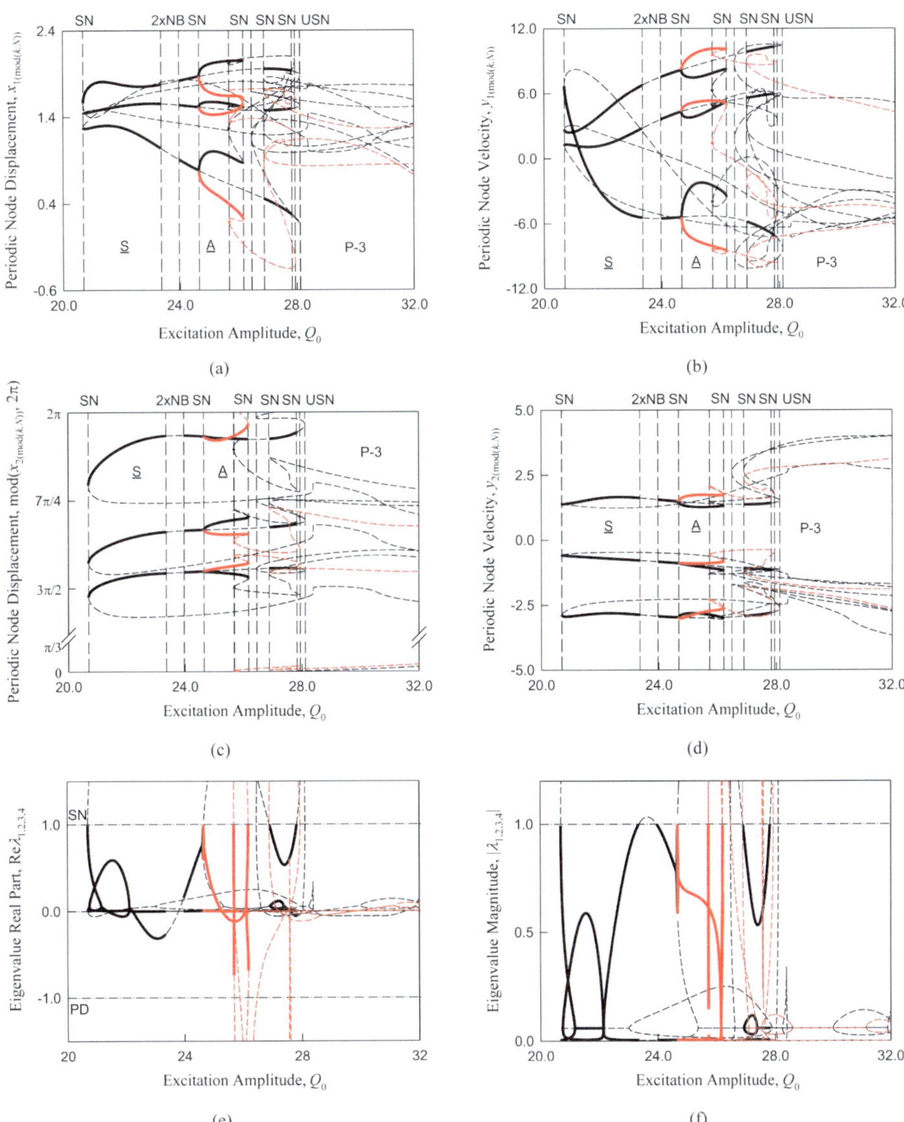

Fig. 6.5 Bifurcation tree of period-3 to chaos varying with excitation amplitude Q_0 (B2). **a** Spring node displacement $x_{1,mod(k,N)}$, **b** spring node velocity $y_{1,mod(k,N)}$, **c** pendulum node displacement $mod(x_{2,mod(k,N)}, 2\pi)$, **d** pendulum node velocity $y_{2,mod(k,N)}$, **e** real part of eigenvalues, **f** eigenvalue magnitudes ($\alpha = 10.0$, $\beta = 20.0$, $\delta = 0.6$, $L = 2.0$, $\Omega = 2.0$)

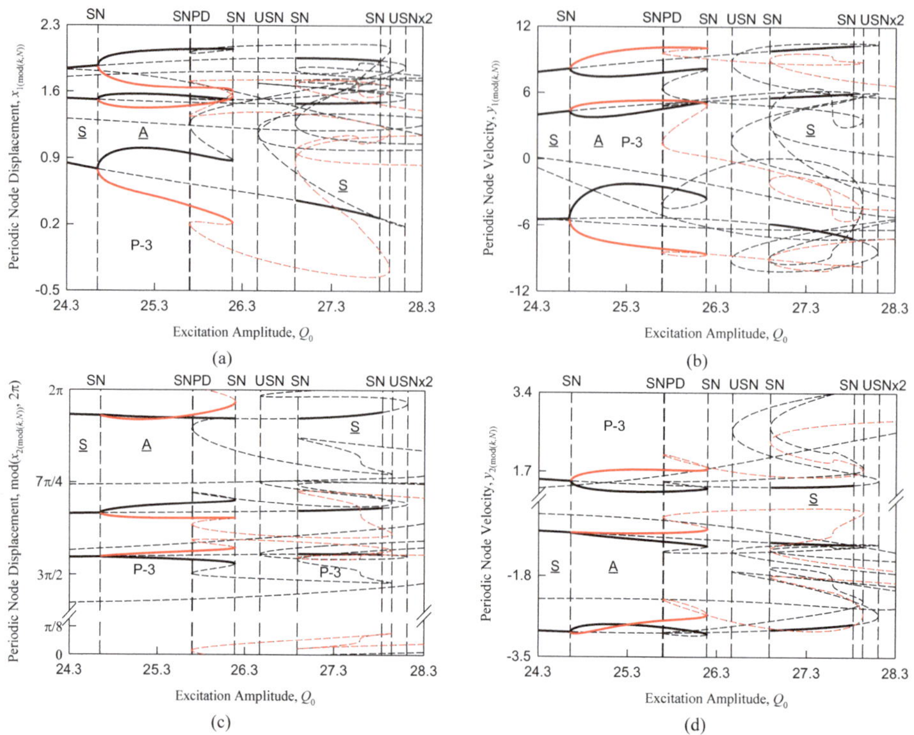

Fig. 6.6 Zoomed view of B2 for $Q_0 \in (24.3, \ 28.3)$. **a** Spring node displacement $x_{1(\text{mod}(k,N))}$, **b** spring node velocity $y_{1(\text{mod}(k,N))}$, **c** pendulum node displacement $\text{mod}(x_{2(\text{mod}(k,N))}, 2\pi)$, **d** pendulum node velocity $y_{2(\text{mod}(k,N))}$ ($\alpha = 10.0$, $\delta = 0.6$, $L = 2.0$, $\Omega = 2.0$)

period-doubling bifurcations take place. As the excitation amplitude Q_0 increases, the introduced unstable period-6 motions come across a pair of unstable-saddle-node bifurcations at $Q_0 \approx 26.2644$ and 25.2597. Such unstable-saddle-node bifurcations are both associated with jumping phenomenon from unstable to unstable. The second branch of the asymmetric period-6 motions in $Q_0 \in (17.5863, \ 18.5290)$ are completely stable. At both $Q_0 \approx 17.5863$ and 18.5290, the period-6 motions have saddle-node bifurcations, which also correspond to the period-doubling bifurcations of period-3 motions. Finally, on the third branch asymmetric period-6 motions exist with two stable ranges: $Q_0 \in (20.3040, \ 20.3354)$ and $(21.3644, 21.4880)$. The two stable ranges are connected through asymmetric unstable period-6 motions. At $Q_0 \approx 20.3040$ and 21.4880, the saddle-node bifurcations happen for the period-6 motions. Such saddle-node bifurcations also correspond to the period-doubling bifurcations of the asymmetric period-3 motions. At $Q_0 \approx 20.3354$ and 21.3644, the asymmetric period-6 motions encounter period-doubling bifurcations, where the period-6 motions are from stable to unstable and further possible cascaded period-doubling bifurcations are introduced.

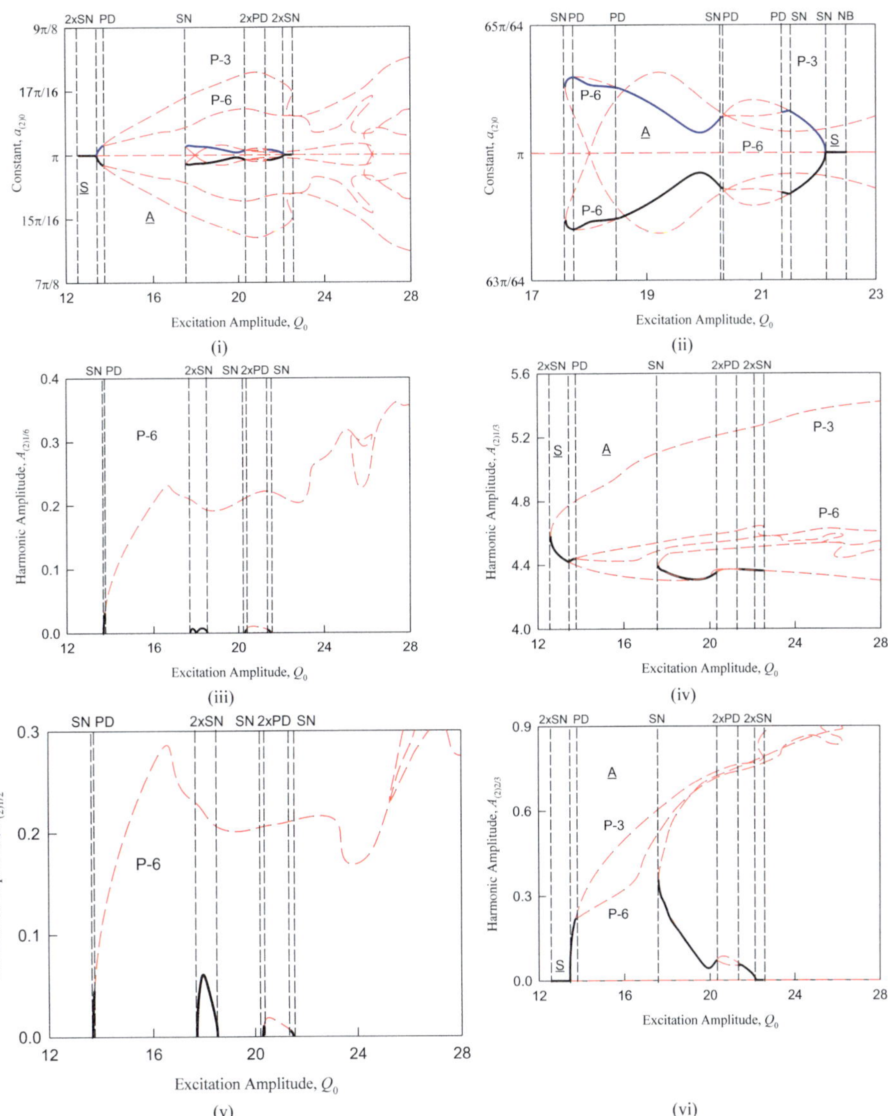

Fig. 6.7 Harmonic amplitude characteristics for bifurcation trees (B1) of period-3 to period-6 motions: **i** $a_{(2)0}$, **ii** zoomed view of $a_{(2)0}$ for $Q_0 \in (17.0, \ 23.0)$. **iii–xvi** $A_{(2)k/m}$ ($m = 6$, $k = 1, 2, 3, \ldots, 6; 8, 10, 12, 18, 24, 30, 120, 360$) ($\beta = 20.0$, $\delta = 0.6$, $L = 2.0$, $\Omega = 2.0$)

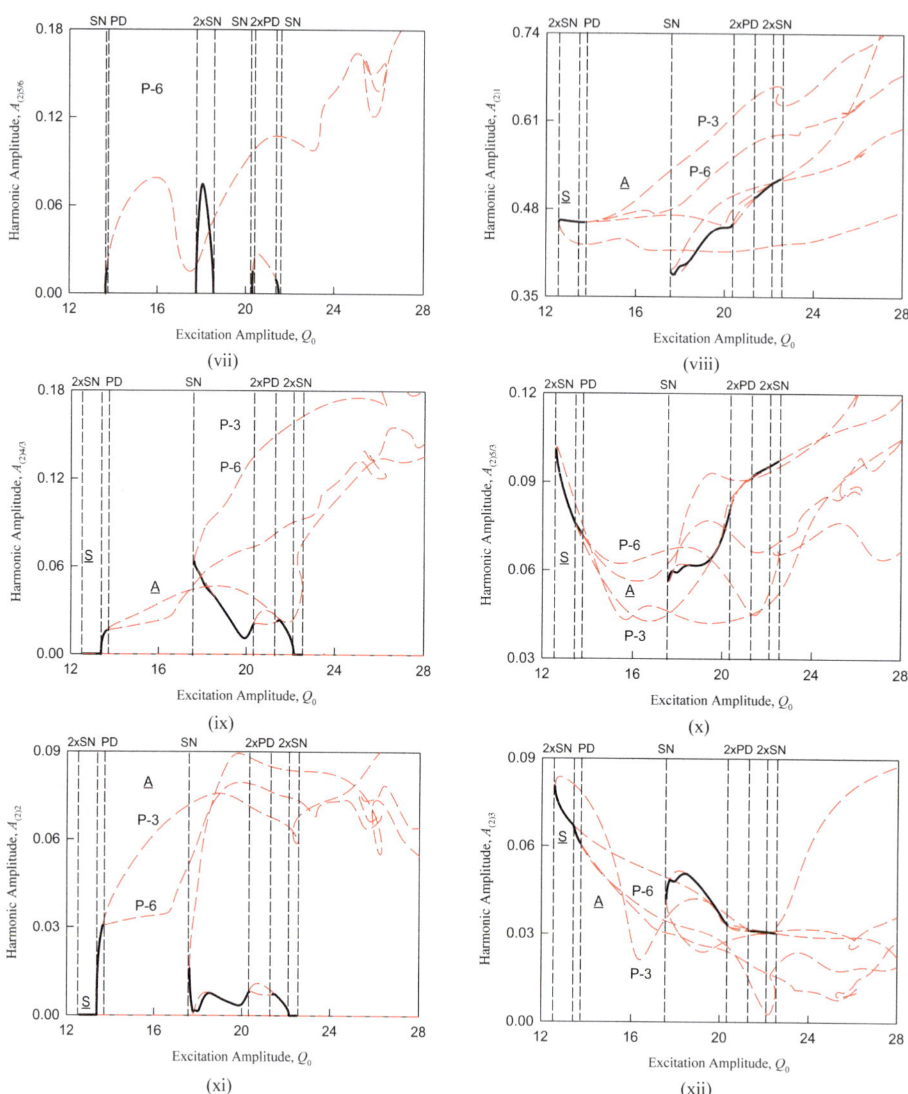

Fig. 6.7 (continued)

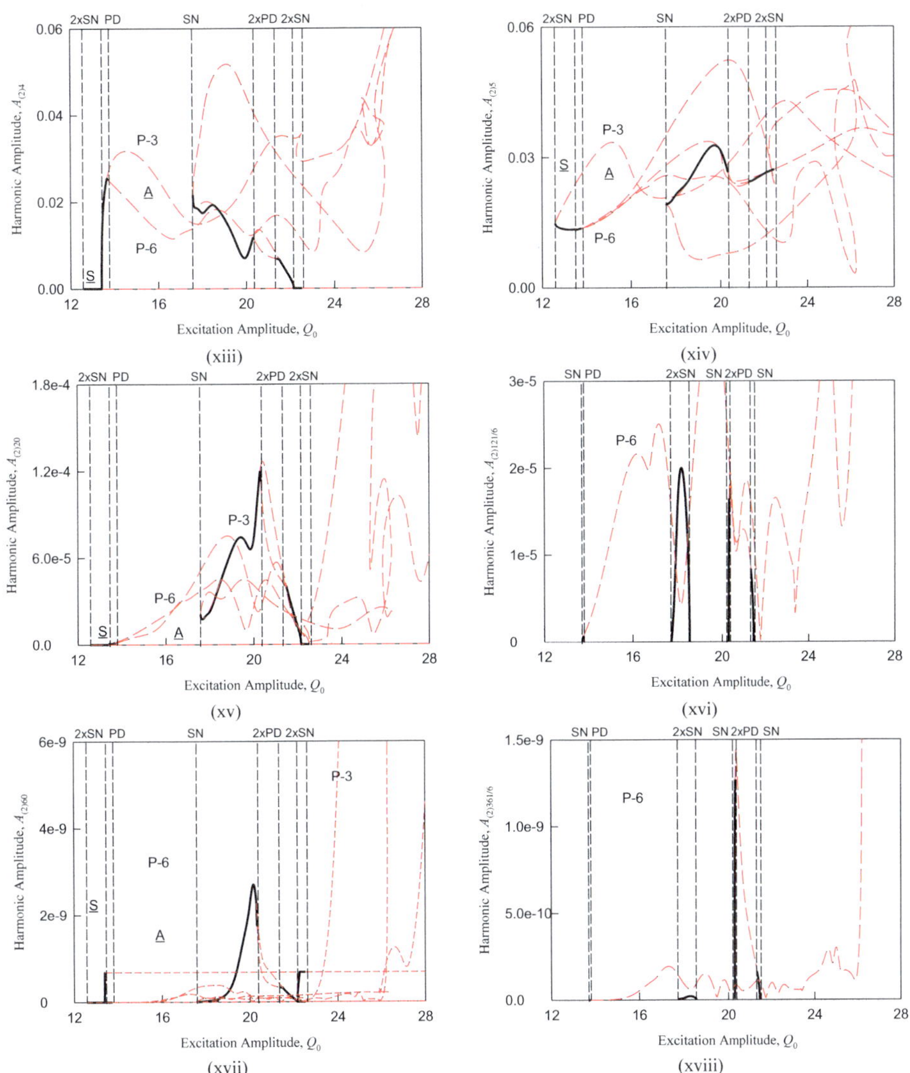

Fig. 6.7 (continued)

The second branch of bifurcation trees (B2) is presented in Fig. 6.5 using similar patterns. For a better illustration, the zoomed detailed views of such bifurcation trees are shown in Fig. 6.6 for $Q_0 \in (24.3, 28.3)$. For such bifurcation trees, the symmetric period-3 motion is stable for three different ranges: $Q_0 \in (20.6879, 23.3600), (23.9700, 24.6610)$, and $(26.8867, 27.8270)$. At $Q_0 \approx 20.6879$, a saddle-saddle bifurcation takes place with a jumping phenomenon from stable to unstable. As the excitation amplitude increases, at $Q_0 \approx 23.3600$, a Neimark bifurcation occurs, where the period-3 motion is from stable to unstable. The unstable symmetric period-3 motion connects to another Neimark bifurcation at $Q_0 \approx 23.9700$, where the motion is from unstable to stable. Then the stable symmetric period-3 motion comes across a saddle node bifurcation at $Q_0 \approx 24.6610$. Such saddle-node bifurcation is from stable to unstable, and a pair of asymmetric period-3 motions are introduced. The associated unstable symmetric period-3 motion connects to another saddle-node bifurcation at $Q_0 \approx 26.8867$, where the motions is from unstable to stable. Such stable symmetric period-3 motion further encounter a saddle-node bifurcation at $Q_0 \approx 27.8270$ and becomes unstable. A pair of unstable saddle-node bifurcations appears for such unstable symmetric period-3 motion at $Q_0 \approx 28.085$ and $Q_0 \approx 26.4679$. Such unstable-saddle-node bifurcations are associated with jumping phenomenon from unstable to unstable.

For asymmetric period-3 motions, two stable ranges are $Q_0 \in (24.6610, 26.887)$ and $(25.6968, 25.7034)$. Such asymmetric period-3 motions are introduced through the period-doubling bifurcations of the symmetric period-3 motion at $Q_0 \approx 24.6610$. The two stable ranges are connected through unstable motions associated with jumping phenomenon. At $Q_0 \approx 24.6610$, saddle-node bifurcations take place for the asymmetric period-3 motions. Such saddle-node bifurcations correspond to the period-doubling bifurcations of the symmetric period-3 motion. As the excitation amplitude increases, saddle-node bifurcations associated with jumping phenomenon occur at $Q_0 \approx 26.1887$. Such saddle-node bifurcations cause the asymmetric period-3 motions to switch from stable to unstable. The introduced unstable motions connects to the second stable range of $Q_0 \in (25.6968, 25.7034)$ through the saddle-node bifurcations at $Q_0 \approx 25.6968$. Finally, at $Q_0 \approx 25.7034$, the asymmetric period-3 motions switches from stable to unstable through the corresponding saddle-node bifurcations. The unstable asymmetric period-3 motions continue developing as the excitation amplitude Q_0 increases. Such an unstable motion further possesses a pair of unstable-saddle-node bifurcations associated with jumping phenomenon at $Q_0 \approx 27.9359$ and 26.8867. Such unstable-saddle-node bifurcations are connected with unstable period-3 motions; and the jumping phenomenon are from unstable to unstable.

6.2   Harmonic Amplitudes

In this section, the bifurcation trees of period-3 motions to chaos will be illustrated through the harmonic amplitudes. Figures 6.7 and 6.8 present harmonic amplitudes for the bifurcation trees (B1) and (B2), respectively. For the bifurcation trees (B1), both period-3 and period-6 motions are illustrated. For the bifurcation trees (B2), only period-3 motions are presented due to the very small stable ranges of the period-6 motions. The acronyms 'SN', 'PD', 'NM', and 'USN' are used for the saddle-node, period-doubling, Neimark, and unstable-saddle-node bifurcations, respectively. The unstable and stable motions are represented by dashed curves and solid curves, respectively. The symmetric and asymmetric motions are indicated by '$\underline{S}$' and '$\underline{A}$', respectively. The stable pairs of coexisting asymmetric motions are illustrated through blue and black colors, respectively. Finally, the period-m motions are labeled as 'P-m' with $m = 3, 6$ in all plots.

The harmonic amplitudes corresponding to the bifurcation trees (B1) are presented in Fig. 6.7 with the given parameters $\alpha = 10.0$, $\beta = 20.0$, $\delta = 0.6$, $L = 2.0$, $\Omega = 2.0$. The constant term $a_{(2)0}$ is presented in Fig. 6.7i for the solution center at $\mathrm{mod}(a_{(2)0}, l\pi) = 0$ ($l = 0, 1, 2, \ldots$). A zoomed detail view is presented in Fig. 6.7ii. The bifurcation tree can be clearly observed. For symmetric motions, $\mathrm{mod}(a_{(2)0}, l\pi) = 0$ with $l = 0, 1, 2, \ldots$. For the asymmetric period-m motion center on the left side of $\mathrm{mod}(a_{(2)0}, l\pi) = 0$, $\mathrm{mod}(l\pi - a_{(2)0}^{L}, 2\pi) = \mathrm{mod}(a_{(2)0}^{R} - l\pi, 2\pi)$ ($l = 0, 1, 2, \ldots$). For the symmetric period-3 motion to induce a pair of asymmetric period-3 motions, a saddle-node bifurcation will occur. For such saddle-node bifurcations, the stable asymmetric period-3 motions appear while the symmetric period-3 motion switches from stable to unstable or from unstable to stable. When jumping phenomenon occurs for stable motions, the saddle-node bifurcations associated to the jumping points take place. Such saddle-node bifurcations causes the motion to switch from stable to unstable. When the jumping phenomenon occurs for unstable motions, the unstable-saddle-node bifurcation corresponding to the jumping points takes place. Such an unstable-saddle-node bifurcation is from unstable to unstable. When the asymmetric period-3 motion experiences a period-doubling bifurcation, the period-6 motions will appear while the asymmetric period-3 motions switch from stable to unstable. On the other hand, such period-doubling bifurcations of the asymmetric period-3 motions also correspond to the saddle-node bifurcations of the period-6 motion. When the period-6 motions possess further cascaded period-doubling bifurcation, possible period-12, 24, ... motions appear and the period-6 motions switch from stable to unstable. Due to the very tiny stable range of such period-12 to period-24, ... motions, they will not be illustrated here. In Fig. 6.7iii, harmonic amplitude $A_{(2)1/6}$ is presented for period-6 motions. For period-3 motions, $A_{(2)1/6} = 0$. A total of three branches of period-6 motions can be clearly observed with their bifurcation points. On the first branch, the bifurcation points include a saddle-node bifurcation, a period-doubling bifurcation, and a pair of unstable-saddle-node bifurcations associated with jumping phenomenon. On the second branch, there is a pair of saddle-node bifurcations. On the third

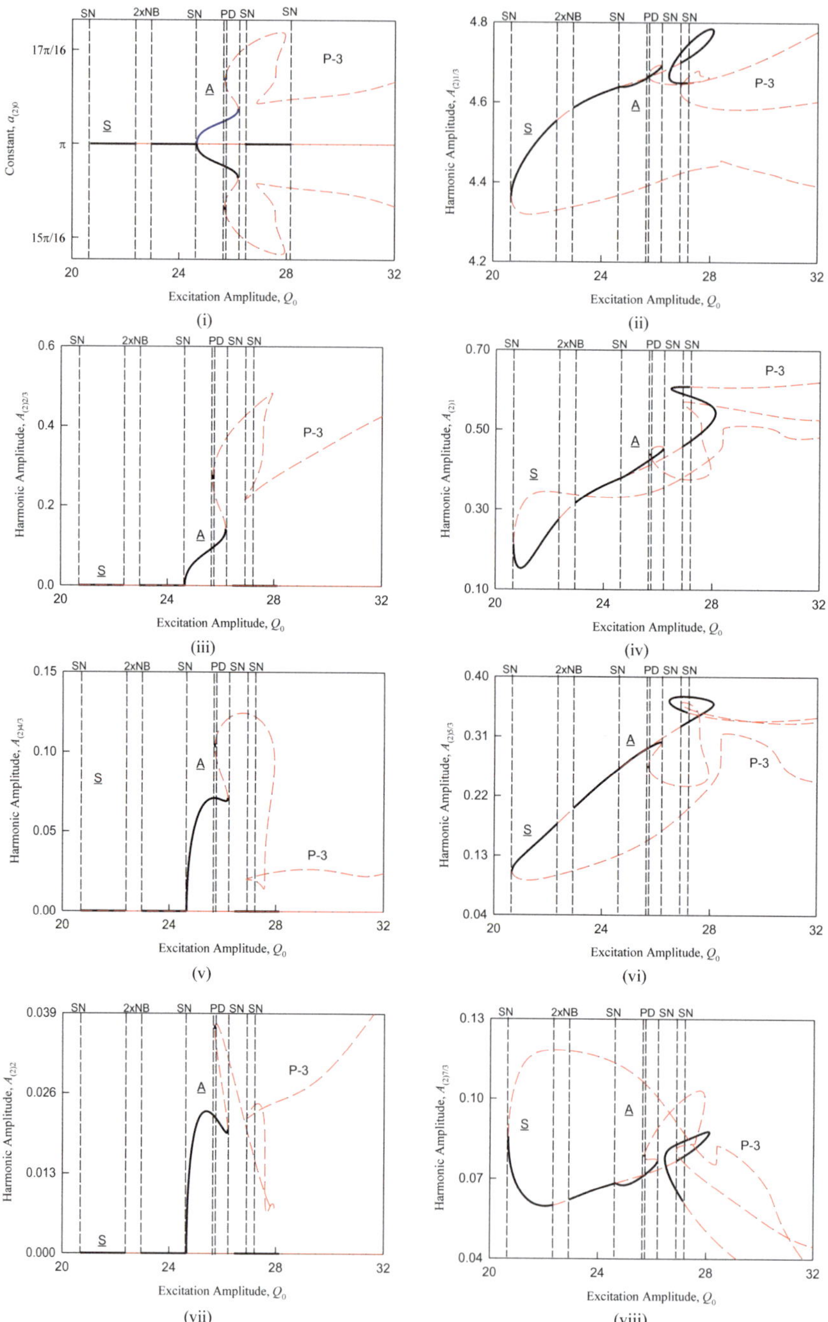

Fig. 6.8 Harmonic amplitude characteristics for bifurcation trees (B2) of period-3 to chaos: **i** $a_{(2)0}$. **ii–xvi** $A_{(2)k/m}$ ($m = 3$, $k = 1, 2, 3, \ldots, 9$; $12, 15, 60, 61, 180, 181$) ($\alpha = 10.0$, $\beta = 20.0$, $\delta = 0.6$, $L = 2.0$, $\Omega = 2.0$)

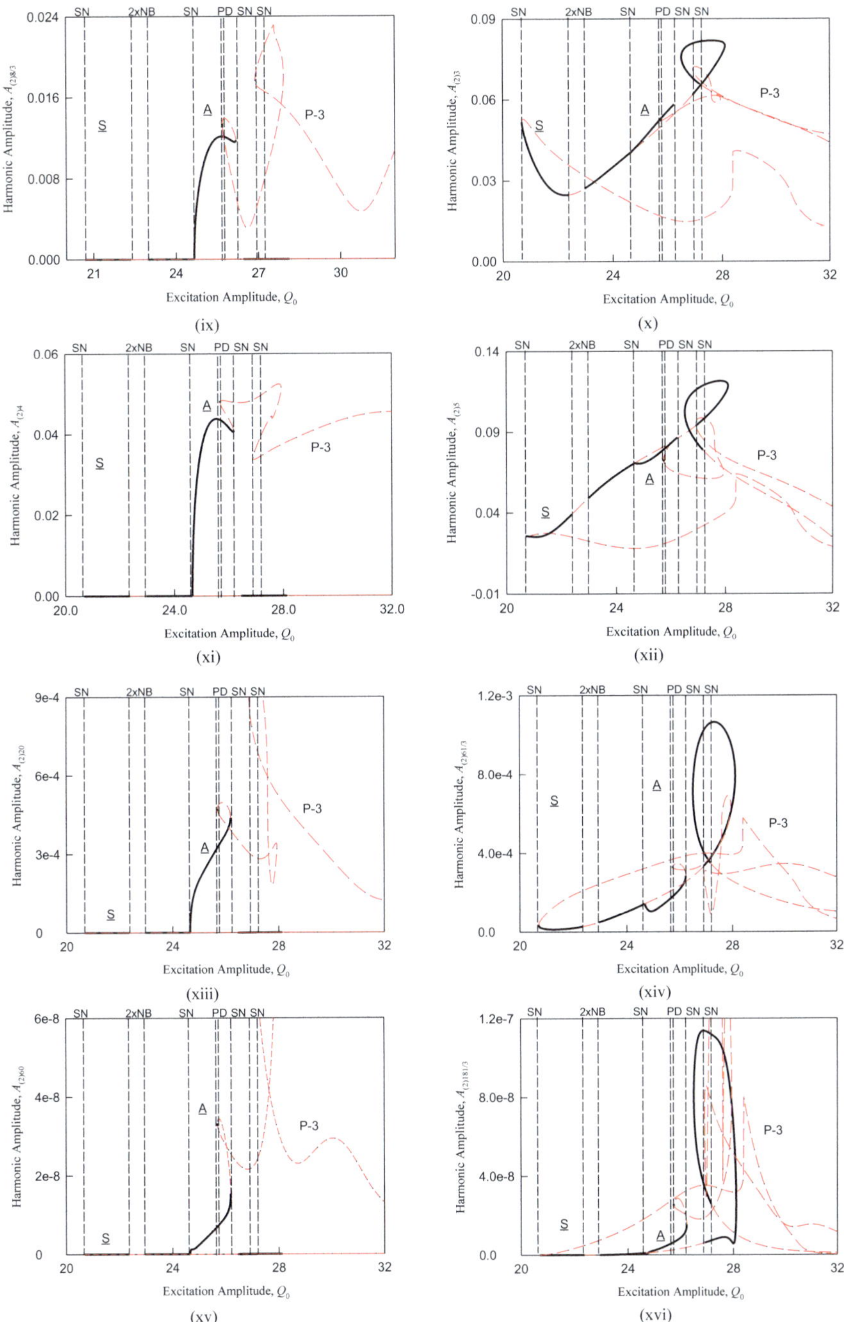

Fig. 6.8 (continued)

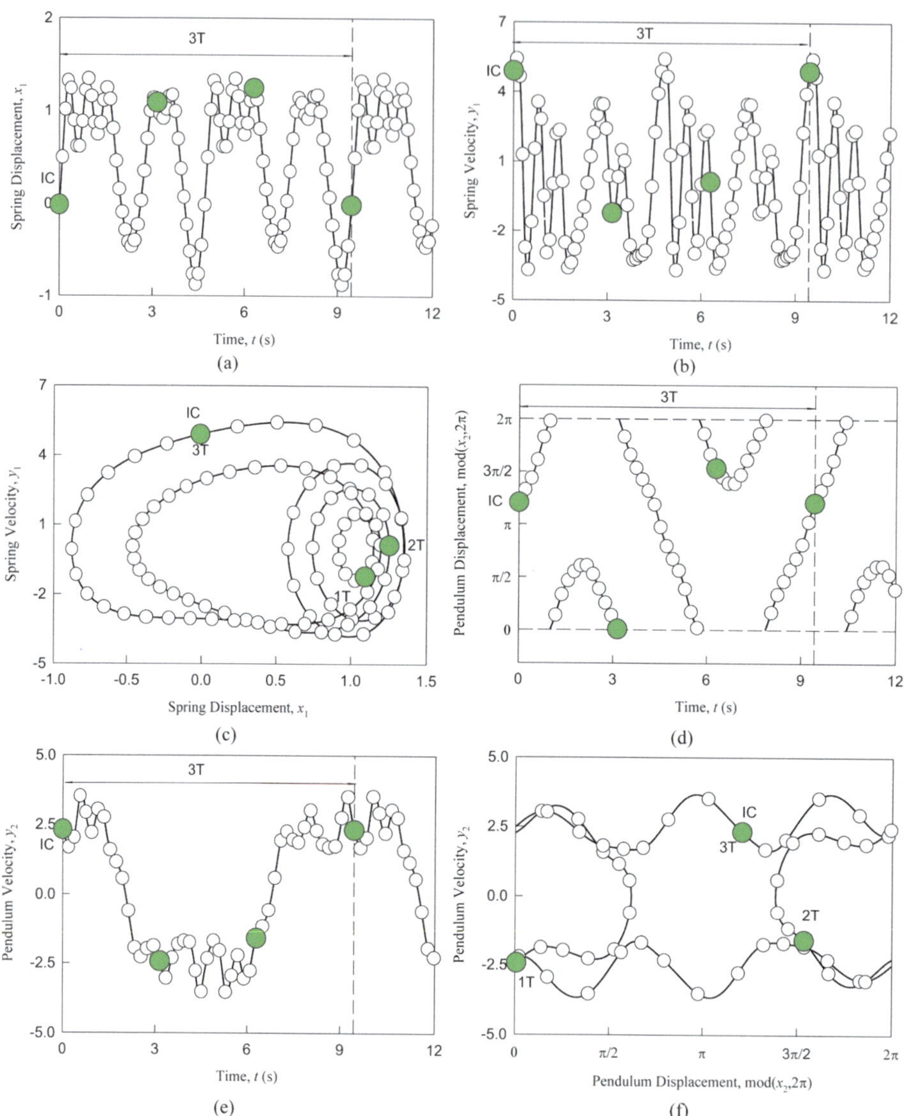

Fig. 6.9 Symmetric period-3 motion at $Q_0 = 12.7$. **a** Spring displacement, **b** spring velocity, **c** spring trajectory, **d** pendulum displacement, **e** pendulum velocity, **f** pendulum trajectory, **g** harmonic amplitude of pendulum displacement, **h** harmonic amplitude of spring displacement, **i** harmonic phase of pendulum displacement, **j** harmonic phase of spring displacement. Initial condition: $t_0 = 0.0$, $x_{1(0)} \approx -0.0122$, $\dot{x}_{1(0)} \approx 4.9088$, $x_{2(0)} \approx 3.7892$, $\dot{x}_{2(0)} \approx 2.3242$. Parameters: ($\alpha = 10.0$, $\beta = 20.0$, $\delta = 0.6$, $L = 2.0$, $\Omega = 2.0$)

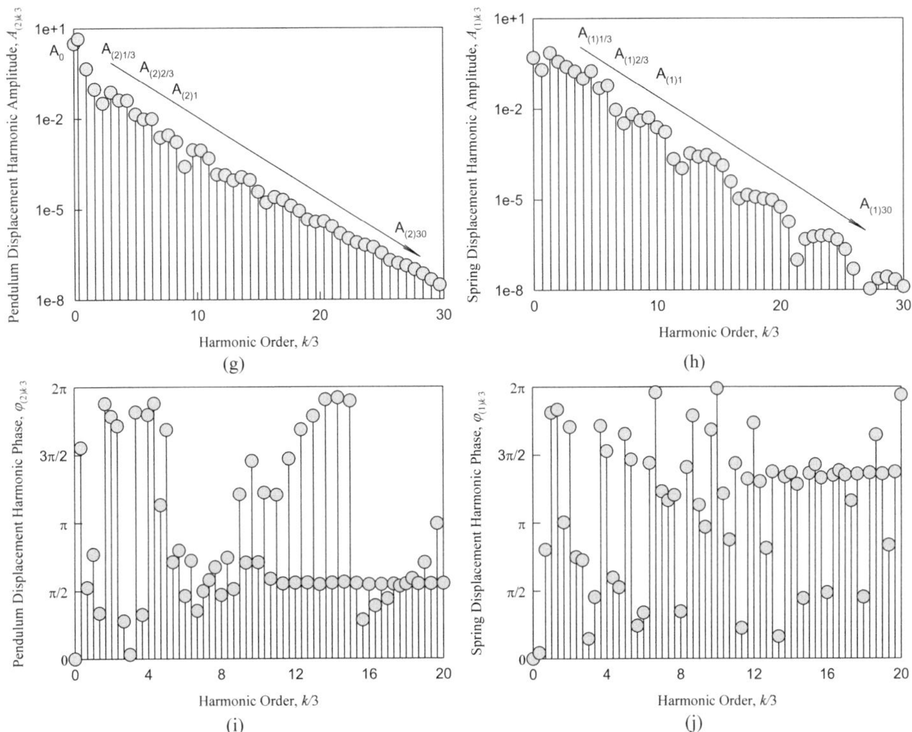

Fig. 6.9 (continued)

branch, the bifurcation points are a pair of saddle-node bifurcations and a pair of period-doubling bifurcations. The quantity level of the harmonic amplitude for the period-6 motions is $A_{(2)1/6} \sim 0.4$. In Fig. 6.7v, vii, the harmonic amplitudes $A_{(2)1/2}$ and $A_{(2)5/6}$ are illustrated similar to $A_{(2)1/6}$ for period-6 motions. Such harmonic amplitudes possess similar bifurcation structure. Both the harmonic amplitudes of $A_{(2)1/2}$ and $A_{(2)5/6}$ equal to zero for period-3 motions. The corresponding quantity levels are $A_{(2)1/2} \sim 0.4$ and $A_{(2)5/6} \sim 0.3$ for the period-6 motions. In Fig. 6.7iv, the harmonic amplitude $A_{(2)1/3}$ is presented for period-3 and period-6 motions. The complete bifurcation tree for both period-3 and period-6 motions can be clearly observed. For the symmetric period-3 motions, the bifurcation points include a saddle-node bifurcation with a jumping phenomenon, three saddle-node bifurcations with asymmetric motions, two Neimark bifurcations, and an unstable-saddle-node bifurcation with asymmetric motions. For the asymmetric period-3 motions, the bifurcation points include: a saddle-node bifurcation with jumping phenomena, five period-doubling bifurcations, and three pairs of unstable-saddle-nodes bifurcations with jumping phenomena. Finally, for the asymmetric period-6 motions, there are three period-doubling bifurcations, five saddle-node bifurcations, and a

Table 6.1   Bifurcations of Bifurcation Trees $\alpha = 10.0$, $\beta = 20.0$, $\delta = 0.6$, $L = 2.0$, $\Omega = 2.0$

Branch 1			Branch 2		
Q_0	Bifurcations	PO	Q_0	Bifurcations	PO
12.5844	SN (J)	P-3	20.6879	SN(J)	P3
13.4250	SN (A)	P-3	23.3600	NB	P-3
13.6787	PD	P-3	23.9700	NB	P-3
13.7400	PD	P-6	24.6610	SN(A)	P-3
22.5087	USN (J)	P-3	26.1887	SN(J)	P-3
22.2753	USN (J)	P-3	25.7034	SN(A)	P-3
26.2644	USN(J)	P-6	25.6968	SN(J)	P-3
25.2597	USN(J)	P-6	27.9359	SN(J)	P-3
32.5115	USN(J)	P-3	26.8867	SN(J)	P-3
32.3473	USN(J)	P-3	26.8867	SN(A)	P-3
25.8532	USN(J)	P-3	27.8270	SN(A)	P-3
26.2918	USN(J)	P-3	28.0850	USN(J)	P-3
17.5863	SN(J)	P-3	26.4679	USN(J)	P-3
17.7300	PD	P-3			
18.5290	PD	P-3			
20.3040	PD	P-3			
20.3354	PD	P-6			
21.3644	PD	P-6			
21.4880	PD	P-3			
22.1400	SN(A)	P-3			
22.5100	NB	P-3			
28.9022	SN(A)	P-3			
28.9100	NB	P-3			
33.0800	SN(A)	P-3			

SN: saddle-node, PD: period-doubling, USN: unstable saddle-node, NB: Neimark bifurcation, A: Asymmetric, J: jumping, P-3: period-3 motion, P-6: peirod-6 motion, PO: periodic orbit

pair of unstable-saddle-node bifurcation associated with jumping phenomenon. The overall quantity level is $A_{(2)1/3} \sim 5.6$. Similar to $A_{(2)1/3}$, the primary harmonic amplitudes $A_{(2)1}$ and the harmonic amplitude $A_{(2)5/3}$ are presented in Fig. 6.7viii and x, respectively. The bifurcation points can be clearly observed with a structure similar to $A_{(2)1/3}$. The corresponding overall quantity levels are $A_{(2)1} \sim 1.8$ and $A_{(2)5/3} \sim 0.3$ for the period-3 to period-6 motions. In Fig. 6.7vi, the harmonic amplitude $A_{(2)2/3}$ is presented. For symmetric motions, $A_{(2)2/3} \sim 0.0$. For the asymmetric period-3 and period-6 motions, bifurcation points are observed obviously. For the asymmetric period-3 motions, the bifurcation points include a saddle-node bifurcation with a jumping phenomenon, five period-doubling bifurcations, and three pairs of unstable-saddle-node bifurcations with jumping phenomenon. For the asymmetric period-6 motions, the bifurcation points are three period-doubling bifurcations, five saddle-node bifurcations, and a pair of unstable-saddle-node bifurcation associated with jumping phenomena. The overall quantity level is $A_{(2)2/3} \sim 1.2$. In Fig. 6.7ix, harmonic amplitudes $A_{(2)4/3}$ is presented with a similar bifurcation structure to $A_{(2)2/3}$. The overall quantity level is $A_{(2)4/3} \sim 0.4$. To reduce abundant illustrations, harmonic amplitudes $A_{(2)k}$ $(k = 2, 3, 4, 5)$ are presented in Fig. 6.7xi, xii, xiii, and xiv. The bifurcation trees are clearly illustrated while the corresponding quantity levels are different. The structures of bifurcation trees of $A_{(2)2}$ and $A_{(2)4}$ are similar to $A_{(2)2/3}$. On the other hand, the bifurcation tree structures of $A_{(2)3}$ and $A_{(2)5}$ are similar to $A_{(2)1/3}$. The overall quantity levels are given as $A_{(2)2} \sim 0.12$, $A_{(2)3} \sim 0.1$, $A_{(2)4} \sim 0.1$, $A_{(2)5} \sim 0.1$. Finally, the harmonic amplitudes of $A_{(2)20}$, $A_{(2)121/6}$, $A_{(2)60}$, and $A_{(2)361/6}$ are presented in Fig. 6.7xv–xviii, respectively. For the harmonic amplitudes $A_{(2)20}$ and $A_{(2)60}$, the complete bifurcation tree structures for both period-3 and period-6 motions are presented; while for the harmonic amplitudes $A_{(2)121/6}$ and $A_{(2)361/6}$, only period-6 motions are observed. The overall maximum quantity levels are $A_{(2)20} \sim 1 \times 10^{-3}$, $A_{(2)121/6} \sim 8 \times 10^{-4}$, $A_{(2)60} \sim 1.5 \times 10^{-6}$, and $A_{(2)361/6} \sim 1.5 \times 10^{-6}$. From the above discussions, about 360 terms $(A_{(2)60})$ are needed for a good approximation of analytical solutions.

The harmonic amplitudes corresponding to the bifurcation trees (B2) are presented in Fig. 6.8 with the same parameters using similar patterns. In Fig. 6.8i, the constant term $a_{(2)0}$ is presented for the solution center at $\mathrm{mod}(a_{(2)0}, l\pi) = 0$ $(l = 0, 1, 2, \ldots)$. For symmetric motions, $\mathrm{mod}(a_{2(0)}, l\pi) = 0$ with $l = 0, 1, 2, \ldots$. For the asymmetric period-m motion center on the left side of $\mathrm{mod}(a_{(2)0}, l\pi) = 0$, $\mathrm{mod}(l\pi - a_{(2)0}^{L}, 2\pi) = \mathrm{mod}(a_{(2)0}^{R} - l\pi, 2\pi)$ $(l = 0, 1, 2, \ldots)$. In Fig. 6.8ii, the harmonic amplitude $A_{(2)1/3}$ is presented. The complete bifurcation tree can be clearly observed. For the symmetric period-3 motions, there are three different stable ranges. The bifurcation points include a saddle-node bifurcation with a jumping phenomenon, a pair of Neimark bifurcations, three saddle-node bifurcations with asymmetric motions, and a pair of unstable-saddle-node bifurcations with jumping phenomena. For the asymmetric period-3 motions, there are two different stable ranges. The bifurcation points include: a saddle-node bifurcation associated with symmetric motions, a pair of saddle-node bifurcations with jumping phenomena, a period-doubling bifurcation, and a pair of unstable-saddle-node bifurcations with jumping phenomena. The overall quantity level is $A_{(2)1/3} \sim 5.0$. With a similar structure of bifurcation trees, the primary harmonic amplitudes $A_{(2)1}$ and the harmonic amplitudes $A_{(2)k/3}$ $(k = 5, 7, 9, 15)$ are presented in Fig. 6.8iv, vi, v, x, and xii, respectively. The bifurcation trees are clearly illustrated. The overall quantity levels of such harmonic amplitudes are $A_{(2)1} \sim 0.7$, $A_{(2)5/3} \sim 0.4$, $A_{(2)7/3} \sim 0.12$, $A_{(2)3} \sim 0.1$, and $A_{(2)5} \sim 0.13$. In Fig. 6.8iii, the harmonic amplitude $A_{(2)2/3}$ is presented. For symmetric motions, $A_{(2)2/3} \sim 0.0$; while for the asymmetric motions, bifurcation points are observed. Such bifurcation points include a saddle-node bifurcation with asymmetric motions, a pair of saddle-node bifurcations associated with jumping phenomena, a period-doubling bifurcation, and a pair of unstable-saddle-node bifurcations with jumping phenomena. The overall quantity level is given as $A_{(2)2/3} \sim 0.6$. In Fig. 6.8v, vii, ix, and xi, the harmonic amplitudes $A_{(2)k/3}$ $(k = 4, 6, 8, 12)$ are presented respectively. For such harmonic amplitudes, the structures of bifurcation trees are similar to $A_{(2)2/3}$. Thus, for symmetric motions, $A_{(2)k/3} \sim 0.0$ $(k = 4, 6, 8, 12)$. For asymmetric motions, the overall quantity levels are $A_{(2)4/3} \sim 0.15$, $A_{(2)2} \sim 0.1$, $A_{(2)8/3} \sim 0.04$, and $A_{(2)4} \sim 0.06$. Finally, the harmonic amplitudes of $A_{(2)20}$, $A_{(2)61/3}$, $A_{(2)60}$, and $A_{(2)181/3}$ are presented in Fig. 6.8xiii–xvi, respectively. For the harmonic amplitudes $A_{(2)20}$ and $A_{(2)60}$, the complete bifurcation tree structures are similar to the one of $A_{(2)2/3}$. For the harmonic amplitudes $A_{(2)61/3}$ and $A_{(2)181/3}$, the bifurcation tree structures are similar to $A_{(2)1/3}$. The overall maximum quantity levels are $A_{(2)20} \sim 1.2 \times 10^{-3}$, $A_{(2)61/3} \sim 1.2 \times 10^{-3}$, $A_{(2)60} \sim 3.0 \times 10^{-7}$, and $A_{(2)181/3} \sim 3.5 \times 10^{-7}$. From the above discussions, about 180 terms $(A_{(2)60})$ are needed for a good approximation in analytical solutions.

6.3 Complex Periodic Motion Illustrations

As in Guo and Luo [1], the numerical simulations of period-3 to period-6 motions will be illustrated in verification to analytical predictions. The parameters of the spring pendulum system are given as $\alpha = 10.0$, $\beta = 20.0$, $\delta = 0.6$, $L = 2.0$, $\Omega = 2.0$. In all illustrations, the solid curves and the hollow circles indicate the numerical and analytical results, respectively. The initial conditions are labelled by the acronym 'IC'. The excitation period is labelled by 'T'. For the chosen parameter of $\Omega = 2$, $T = \pi$. The periodic nodes of symmetric periodic motions are indicated by the green circles. For asymmetric motions, the black and red colors represent the black and red branch motions from the bifurcation trees, respectively. The green and red circles indicate the periodic nodes of the black and red branches, respectively.

In Fig. 6.9, a sampled symmetric period-3 motion is illustrated for $Q_0 = 12.7$. The corresponding initial conditions are $t_0 = 0.0$, $x_{1(0)} \approx -0.0122$, $\dot{x}_{1(0)} \approx 4.9088$, $x_{2(0)} \approx 3.7892$, $\dot{x}_{2(0)} \approx 2.3242$. The displacement, velocity, and trajectory of the spring motion are illustrated in Fig. 6.9a–c, respectively. The displacement, velocity, and trajectory of the pendulum motion are demonstrated in Fig. 6.9d–f, respectively. The period of the spring motion is π, which is exactly one period of the excitation period. On the other hand, the period-3 motion of the pendulum is $3T = 3\pi$. Thus, the overall period of such a periodic motion is $3T = 3\pi$. The spring demonstrates three cycles of exact same motions within the period, while the pendulum completes only one cycle of motion. The displacement of the pendulum always goes from zero to 2π. The pendulum is swinging symmetrically about the point 0 or 2π with one rotation clockwise and another counter-clockwise. The trajectory of the pendulum motion in phase plane is symmetric to itself according to the point $(\pi,\ 0)$, while no symmetry is demonstrated for the trajectory of the spring motion. The trajectories of both the spring and pendulum form a closed loop in phase plane. The harmonic amplitudes are presented in Fig. 6.9g, h for the pendulum and spring displacements, respectively. The harmonic amplitudes drop rapidly to below 10^{-8} within 30 harmonic terms ($A_{(1)30}$ and $A_{(2)30}$), which demonstrates a very good convergence. Finally, the corresponding harmonic phases are illustrated in Fig. 6.9i, j for the pendulum and spring displacements, respectively.

Following the bifurcation trees (B1) from analytical prediction, the symmetric period-3 motion comes across a saddle node bifurcation associated with asymmetric motions at $Q_0 \approx 13.4250$. Such saddle-node bifurcation introduces a pair of co-existing asymmetric period-3 motions. In Fig. 6.10, such a pair of asymmetric period-3 motions is presented for $Q_0 = 13.6$. The initial conditions at $t_0 = 0.0$ are given as $x_{1(0)} \approx 0.3196$, $\dot{x}_{1(0)} \approx 5.9787$, $x_{2(0)} \approx 4.0502$, $\dot{x}_{2(0)} \approx 2.2342$ for the black branch, and $x_{1(0)} \approx 1.2341$, $\dot{x}_{1(0)} \approx -0.6762$, $x_{2(0)} \approx 4.7546$, $\dot{x}_{2(0)} \approx -1.0948$ for the red branch. The displacement, velocity, and trajectory of the spring motions are illustrated in Fig. 6.10a–c, respectively. For both the black and red branches motions, the spring demonstrates exact same motions with a phase difference of $\pi/2$. Such motions conform same trajectories in the phase plane as show in Fig. 6.10c. The period of the spring motions is 3π, which is three times the excitation period. The phase trajectories of the pendulum for the black and red branches are presented in Fig. 6.10d, e, respectively. Unlike the symmetric motion, the asymmetric trajectories are skew-symmetric to each other about $(\pi,\ 0)$ in phase plane. The period of the pendulum motions is $3T = 3\pi$, which is the same with the spring motions. Thus, the overall period of such periodic motions is $3T = 3\pi$. In Fig. 6.10f, g, the harmonic amplitudes of the pendulum and spring displacements are presented, respectively. The harmonic amplitudes of the black and red branches are identical for both the pendulum and spring motions. Such harmonic amplitudes drop rapidly to about 10^{-8} within thirtieth-order (90 harmonic terms), which demonstrate a very good convergence. On the other hand, the corresponding harmonics phases are presented in Fig. 6.10h, i for the pendulum and spring, respectively. Such harmonic phases are different between the black and red branches for such pendulum and spring motions, as shown in Fig. 6.10h, i. The differences of such harmonic phases between the black and red branches are illustrated in Fig. 6.10j, k for the pendulum and spring motions, respectively. For the spring motion the phase difference of the constant term is $\Delta\varphi_{(1)0} = 0$. And $\Delta\varphi_{(1)k} = \mathrm{mod}\big((2 - \tfrac{k}{3})\pi, 2\pi\big)$, $k = 1, 2, 3\ldots$ for harmonic order $k = 1, 2, 3, \ldots$. For the pendulum motion the phase difference of the constant term is $\Delta\varphi_{(2)0} = 0$. The corresponding phase differences are given as $\Delta\varphi_{(2)k} = \mathrm{mod}\big((1 + \tfrac{k}{3})\pi, 2\pi\big)$, $k = 1, 2, 3\ldots$ for harmonic order $k = 1, 2, 3, \ldots$.

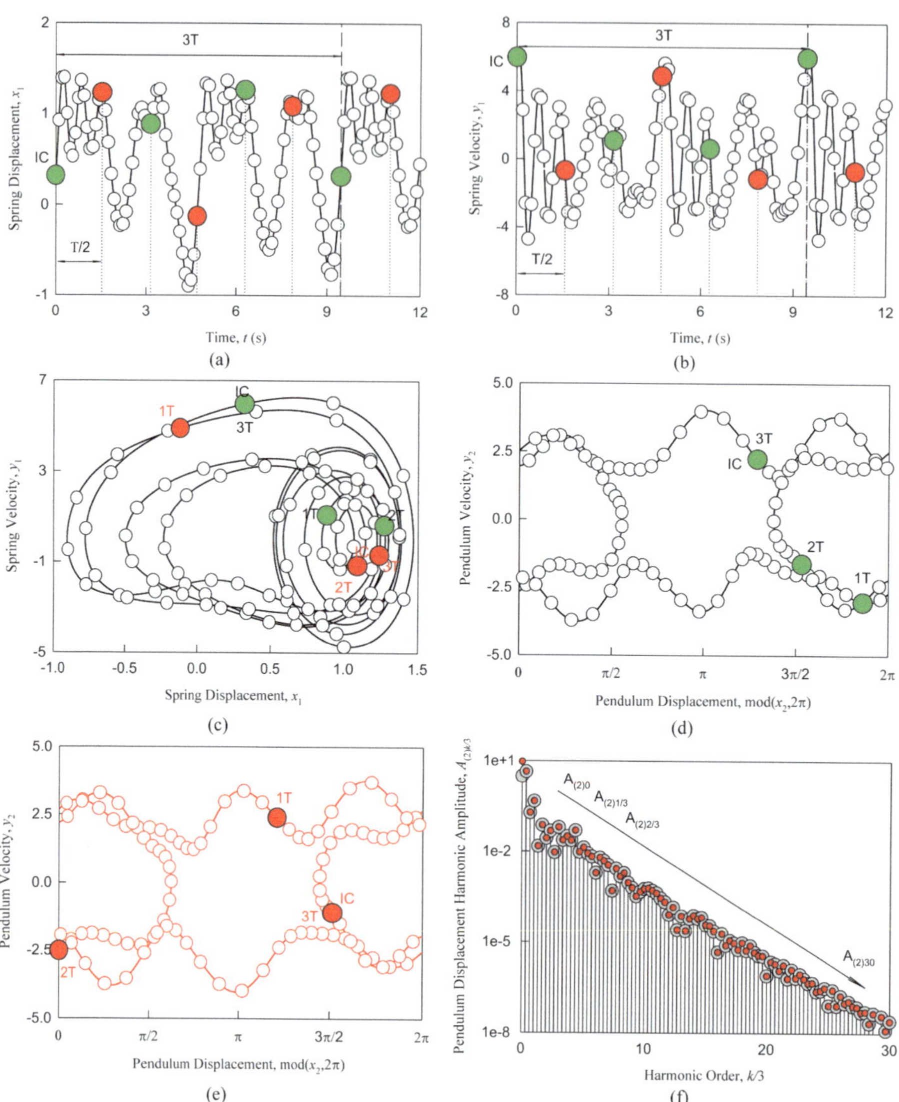

Fig. 6.10 Asymmetric period-3 motion at $Q_0 = 13.6$. Spring: **a** displacement, **b** velocity, **c** trajectory; pendulum: **d** black branch trajectory, **e** red branch trajectory; **f** pendulum harmonic amplitudes, **g** spring harmonic amplitudes, **h** pendulum harmonic phases, **i** spring harmonic phases, **j** pendulum harmonics phase differences, **k** spring harmonic phase differences. Initial condition: $t_0 = 0.0$, black branch: $x_{1(0)} \approx 0.3196$, $\dot{x}_{1(0)} \approx 5.9787$, $x_{2(0)} \approx 4.0502$, $\dot{x}_{2(0)} \approx 2.2342$; red branch: $x_{1(0)} \approx 1.2341$, $\dot{x}_{1(0)} \approx -0.6762$, $x_{2(0)} \approx 4.7546$, $\dot{x}_{2(0)} \approx -1.0948$. Parameters: ($\alpha = 10.0$, $\beta = 20.0$, $\delta = 0.6$, $L = 2.0$, $\Omega = 2.0$)

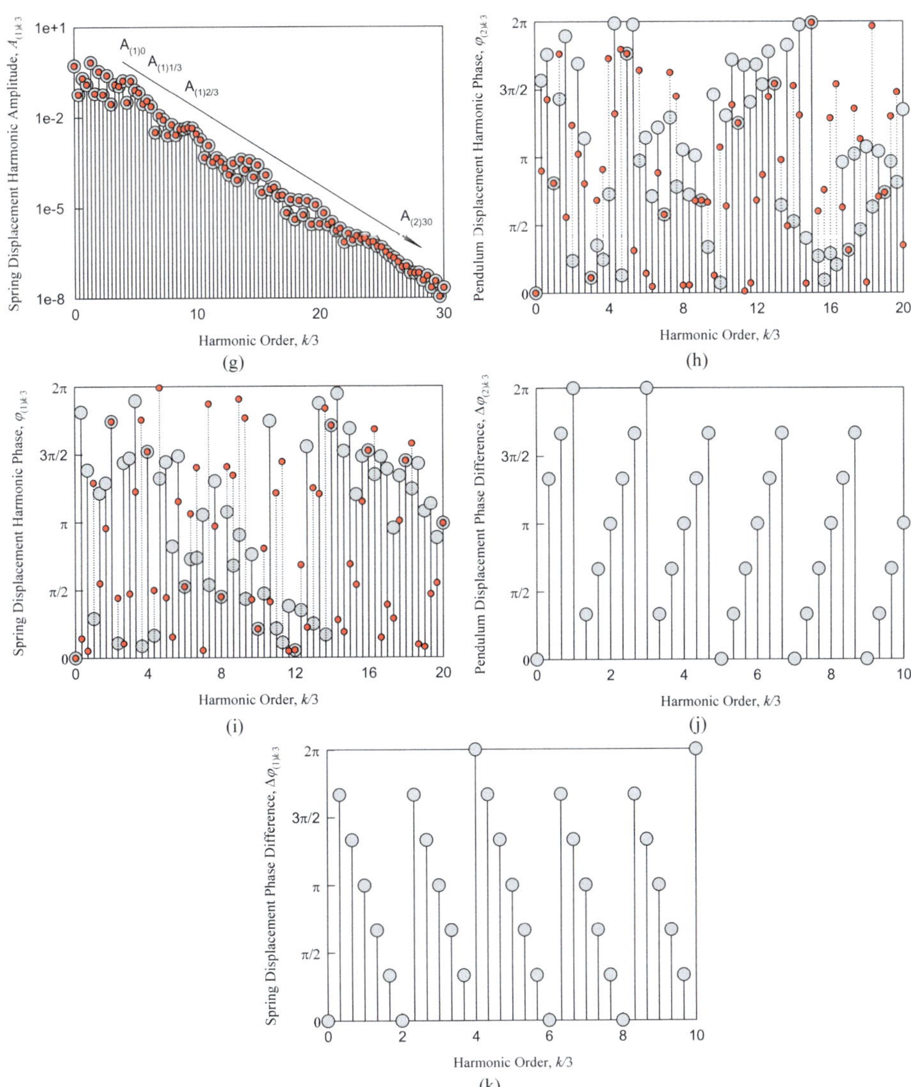

Fig. 6.10 (continued)

Following the development of the bifurcation tree (B1), the asymmetric period-3 motions encounter period-doubling bifurcations at $Q_0 \approx 13.6787$. Such a period-doubling bifurcation introduce a pair of asymmetric period-6 motions. A pair of such asymmetric period-6 motions is demonstrated for $Q_0 = 13.73$ in Fig. 6.11. The initial condition for the black branch motion is $t_0 = 0.0$, $x_{1(0)} \approx 1.2796$, $\dot{x}_{1(0)} \approx 0.6843$, $x_{2(0)} \approx 4.8186$, and $\dot{x}_{2(0)} \approx -1.7342$, while for the red branch, the initial condition is $t_0 = 0.0$, $x_{1(0)} \approx -0.1810$, $\dot{x}_{1(0)} \approx 4.7664$, $x_{2(0)} \approx 3.7328$, and $\dot{x}_{2(0)} \approx 2.4011$. The displacement, velocity, and trajectory of the spring motions for such asymmetric coexisting motions are illustrated in Fig. 6.11a–c, respectively. Similar to the asymmetric period-3 motions, the spring demonstrates identical motions for the black and red branches with a phase difference of $\pi/2$. Such spring motions conform the same trajectories in phase plane for the black and red branches, as shown in Fig. 6.11c. The trajectories of the pendulum motions from the black and red branches are demonstrated in phase plane in Fig. 6.11d, e, respectively. Similar to the asymmetric period-3 motions, such trajectories of the pendulum motions are skew-symmetric to each other according to the point $(\pi, 0)$ in phase plane. In Fig. 6.11f, g, the harmonic amplitudes of the pendulum and spring displacements are presented, respectively. The harmonic amplitudes of the black and red branches are identical for both the pendulum and spring motions except for the constant term $a_{(2)0}$. Such harmonic amplitudes drop rapidly to about 10^{-8} within thirtieth-order (180 harmonic terms), which continue to demonstrate a very good convergence. On the other hand, the corresponding harmonics phases are presented in Fig. 6.11h, i for the pendulum and spring displacements, respectively. Such harmonic phases are different between the black and red branches for the pendulum and spring motions. The differences of such harmonic phases between the black and red branches are illustrated in Fig. 6.11j, k for the pendulum and spring motions, respectively. For the spring motions, the phase difference of the constant term is $\Delta\varphi_{(1)0} = 0$. And $\Delta\varphi_{(1)k} = \mathrm{mod}\left(\frac{k}{6}\pi, 2\pi\right)$, $k = 1, 2, 3\ldots$ for harmonic order $k = 1, 2, 3, \ldots$. For the pendulum motions, the phase difference of the constant term is $\Delta\varphi_{(2)0} = 0$. The corresponding phase differences are given as $\Delta\varphi_{(2)k} = \mathrm{mod}\left(\left(1 + \frac{k}{6}\right)\pi, 2\pi\right)$, $k = 1, 2, 3\ldots$ for harmonic order $k = 1, 2, 3, \ldots$.

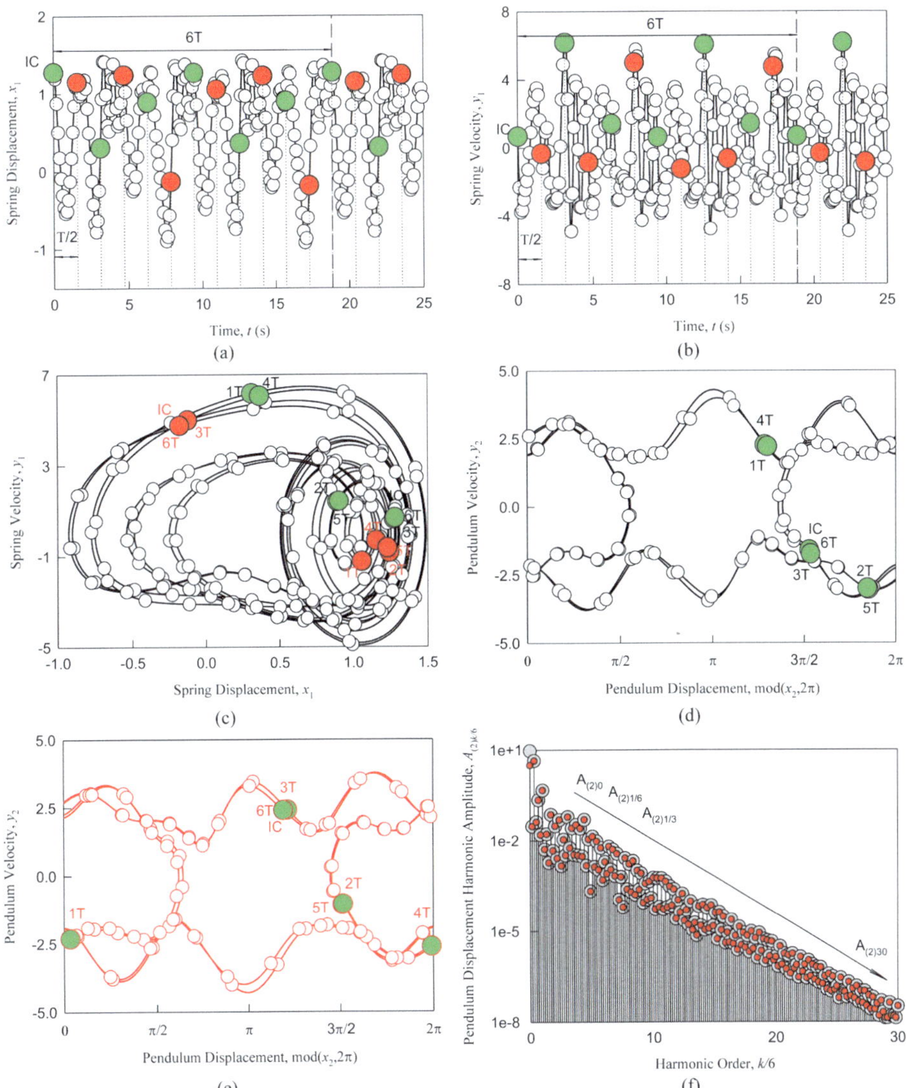

Fig. 6.11 Asymmetric period-6 motion at $Q_0 = 13.73$. Spring: **a** displacement, **b** velocity, **c** trajectory; pendulum: **d** black branch trajectory, **e** red branch trajectory, **f** pendulum harmonic amplitudes, **g** spring harmonic amplitudes, **h** pendulum harmonic phases, **i** spring harmonic phases, **j** pendulum harmonics phase differences, **k** spring harmonic phase differences. Initial condition: $t_0 = 0.0$, black branch: $x_{1(0)} \approx 1.2796$, $\dot{x}_{1(0)} \approx 0.6843$, $x_{2(0)} \approx 4.8186$, $\dot{x}_{2(0)} \approx -1.7342$; red branch: $x_{1(0)} \approx -0.1810$, $\dot{x}_{1(0)} \approx 4.7664$, $x_{2(0)} \approx 3.7328$, $\dot{x}_{2(0)} \approx 2.4011$. Parameters: ($\alpha = 10.0$, $\beta = 20.0$, $\delta = 0.6$, $L = 2.0$, $\Omega = 2.0$)

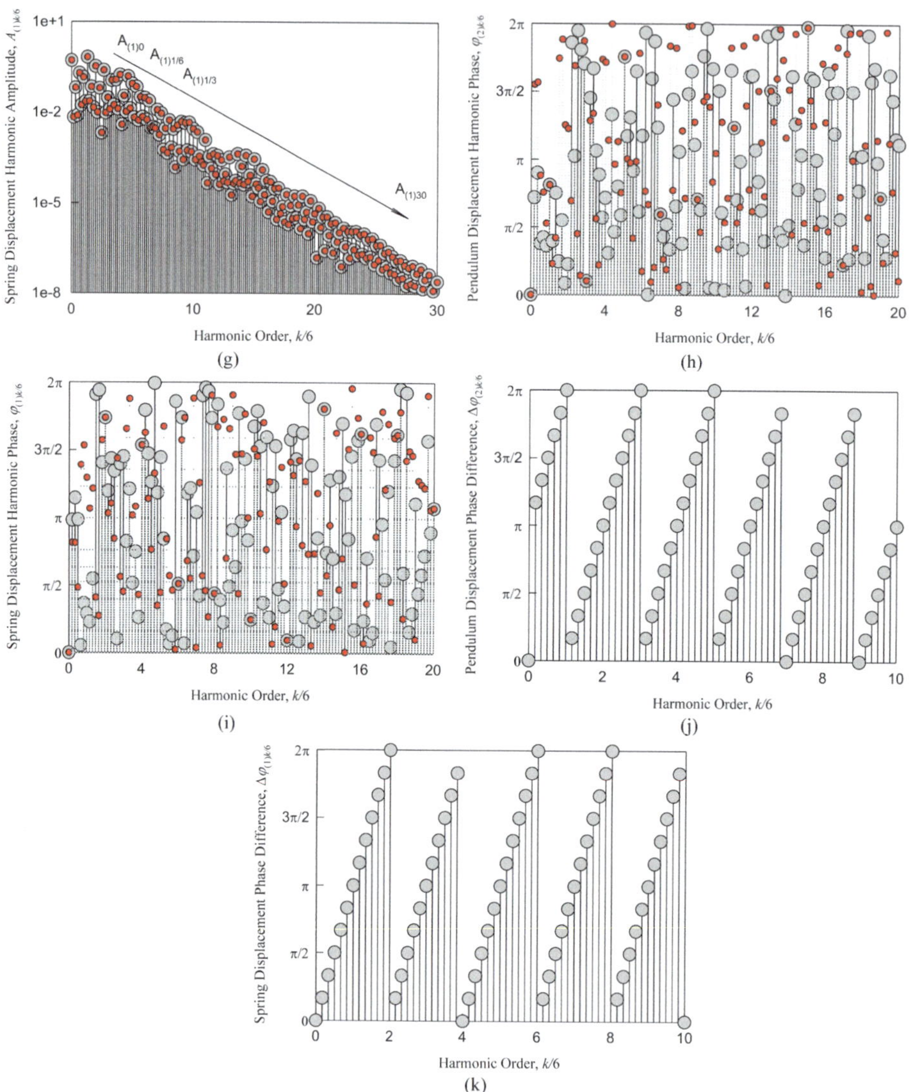

Fig. 6.11 (continued)

Reference

1. Guo Y, Luo ACJ (2022) Period-3 motions to chaos in a periodically forced nonlinear-spring pendulum. *AIP Chao*, **32**, 103129.